One Life
Movement

By

Kimberley Dickinson

An awkward and painful stumble through

one woman's path to finding purpose.

Published By Kimberley Dickinson at Create Space

Copyright 2016 Kimberley Dickinson
All rights reserved

ISBN 978-0-9958400-1-0 (pbk)
ISBN 978-0-9958400-0-3 (eBook)

Library and Archives Canada Cataloguing in Publication

Cover Art by Philip Zu Putlitz, a creative genius

Author Photography by Roxanne Low Photography

Create Space Edition, License Notes

À mon fils Gavin, mon amour, ma lumière, mi vida.
Je suis toujours avec toi, à Neptune et au-delà.

To my parents, siblings and very close friends for
remaining connected with me, despite everything.

To Jill and Mike because #scorpio.

+++++

Table of Contents

PREFACE

I have lived a mighty colourful life up to this point. I have had some very positive times and I have had some very dark times. Every event in my life has served a purpose whether or not I realized it at the time. It has taken me to be completely broken down to figure out what my purpose in life actually is. Painfully I have made the wrong choices and denied myself the ability to grow when I most needed to. I have stubbornly held on to beliefs about myself and my world, not realizing that these beliefs were not my own. They were everything I thought I should be and everything I thought the world wanted me to be, and this has led me to much pain and suffering.

I have met some absolutely lovely people throughout my life that have served the purpose of always trying to propel me towards a higher understanding of who I am as a person. Many of these people have had mental health issues. Unfortunately like many cultural stigmas, there is a strongly held belief that mental wellness, or more appropriately the lack thereof, should not be talked about. People don't feel comfortable talking about anything regarding this in general in the society I grew up in and I have found myself being impacted by it numerous times.

Mental illness is virtually everywhere. I can definitively say I have not known one single human being that has not had experience with mental un-wellness in their life either directly or indirectly. Whether it be depression, eating problems, low self-worth, low self-respect, distorted body image, suicide, bipolar, schizophrenia, post-partum, lost identity, addiction, loneliness, and personality disorders. Virtually everyone I have ever met has at least had some shade of experience with someone (or themselves) dealing with any given one of these topics at some point in their life. And yet, no one talks about it. Friends, family, children, lovers; someone always knows someone who has had a struggle. Just the other day I and another random mom at swimming lessons got to talking and she shared with me how manipulative, controlling and abusive her last boyfriend had been. She actually seemed relieved to be able to discuss that with someone else in an open forum.

There is less and less funding put towards issues related to mental health, and very little in terms of social supports for anyone struggling to cope with it personally or second hand. This doesn't make any sense. For an issue that is so prolific, that literally touches the lives of everyone around and including the individual, to have it be some kind of dismissed entity seems quite frankly backwards. I deal with mental health issues on a daily basis in my professional life and it has definitely helped me to identify, manage and understand just how far barriers go for those who struggle with such issues.

I also deal with mental health issues on a daily basis in my personal life. I have always been around mental health issues in fact, throughout my entire life. We hosted numerous foster children in my youth. These young souls lived and played with members of my family, attended school and events with us and were a part of our every day life. I met the most gorgeous most beautiful souls through fostering. Drug addicted parents, verbally, physically and sexually abusive relatives, mental health disorders, and abandonment. These children had levels of damage that went far deeper than anything most adults would ever have to face, all before the age of 10.

We struggled as a family because so much of our energy went into helping others, but I know in the depths of my soul that we saved countless lives because of it. It remains to be seen the generational pain that people experience because of living through less than ideal home lives. Mental health issues affect every facet of society; the healthcare system, housing, social programs, religious entities, access to basic human needs such as toiletries, social supports, family breakdowns, reciprocal substance misuse and abuse. It is like one massive, carefully constructed, complex and weaving set of dominoes.

And yet no one feels comfortable talking about it. We are taught to not talk about such things. "What will people think?" But in reality, we have all dealt with it, we all know what it is like, but we don't know what we can do to change that. Lately there have been discussions in the media about certain news stories that have cropped up, and I'm beginning to see more dialogue about social injustices, about people opening up and being honest with others. And it is a step in the right direction. The more people are exposed to the idea that something is and should be talked about is part of making the discussion normalized. This in turn will drift us towards building at least part of a solution.

The year 2014 was tough. I lost a student to a drug overdose, a friend to a completed suicide, and was able to intervene very closely in another attempted abrupt end to a life. The last time I saw my student before she passed, she was walking out of my classroom less than a week earlier. She had struggled with addiction and had finally gotten herself off of her chosen elixir. I told her I was proud of her, and she smiled and thanked me. A few days later she was found in a bathroom of a coffee shop. Living through these experiences taught me some lessons about not taking anyone for granted, not failing to be there for someone regardless of how we are connected, that everyone goes through periods of sorrow and loneliness, and that I had to learn how to reach out more and give what I could, even if it wasn't much by my standards. I was terribly saddened by each passing, not because of how they chose to go but sad that they are gone.

I have no doubt each one of these humans felt a deep and profound loneliness, a lack of connection. We have all felt alone at one time or another. Some of us have even been crippled by loneliness, including myself. I see it so often with clients, friends and family members. No one is immune, I have felt it myself and it is not necessary. None of us are truly alone, but somehow along the way we have forgotten we are all human and the vast majority of us feel the feelings. We idolize strangers and each other, and then tear them down when we find that they are only human stumbling through normal human mistakes. We must stop tearing each other down, and we must stop tearing ourselves down when we feel we do not measure up to someone else's version of ourselves.

My very best, most beautiful friends have cried in my arms and me in theirs for feeling that we were truly lost and alone in the vastness of space and time. The loneliest I have ever felt was when I was married, a time when I was actually legally bound to another human being and yet I felt my existence was nowhere to be found. I also dated a very successful professional for quite a while and he was the loneliest human being I have ever known and it felt as though he hated himself. It seemed as if he loathed the very fabric of his own being even though for all intents and purposes he had EVERYTHING anyone could have ever aspired to have: the highest education, an extremely comfortable income, a healthy job, a healthy body, an amazing property, a loving family, an awesome girlfriend. But none of it was enough.

The sorrow in this man as he went about tending to his land, his home, his job and his things was all over his face and in his rigid body. He was full of anger and hate and misery, yet he put all his effort in to not portraying his true self in his daily routine or in his job. He actually believed that no one could see through this, but I did eventually and that's probably why I stuck around for as long as I did. Despite him not being faithful, lying via omission, going through a plentiful arsenal of pharmaceuticals and alcohol, I lingered. Because I knew what people do when they are hurting, because it is not okay to say "I'm feeling really rough today, I could really use some support" instead we say "let's go party!" and we meet up with random online dating atrocities. We have affairs, we eat too much or not enough, we internalize everything and blame others for everything, we watch countless hours of TV and we try to control things for no reason. I identified that in him, and I wanted to sprinkle my fairy dust on him as I always do, except that doesn't work if they do not wish to be sprinkled on. I'm not even sure if he identified then, or will ever be able to release the pain he held in himself, just as most of the men I have found myself with.

I am not pure as the wheat in the field's kind of human being. I am stumbling through this life as we all do. In fact I am a failure, I fail all the time, and usually spectacularly in a blaze of gore and flames. I am really quite impressive at failing, and making bad decisions based on information I received at a given time. I have made poor choices, like sooooo incredibly pooooorrrr, and I have said things that were taken the wrong way. But I am trying with the heat of a thousand suns to do what I can, where I am with what I am given. I understand why I have been given the lessons I have of late. I understand that I am worthy of so much more than average. I am stronger than I ever have been, but that is not to say that I am not fortified. I still cry.

Yesterday I cried when I realized that despite an extremely stressful few weeks at work and personally, I was actually feeling satisfied with myself. I was sitting in the car on the way home, sun in my eyes and my now jammed CD player making mechanical grinding noises in the background, a stench of rotten smoothie coming from somewhere in the car; I felt satiated in knowing that what I am doing is affecting change, all around me and including me. I am being given these small gifts from every interaction I have with friends, family, and business contacts that are all helping me to get to where I am going. I am fighting to advocate for all kinds of people, and it feels good.

The shame we all feel for our failures is deep and heavy and I am not going to be shameful any more. I have accepted myself for who I am, what I have to give, and the experiences I have had. I am satisfied with where I am in my life, and I know with great certainty that I will continue on doing so. There are a thousand things that anyone could rightfully find wrong with me, and there are a million things that someone could find more than right. I have no interest in being around those who want to focus on those thousand things in me, when they have no interest in focusing to ameliorate their own thousand things. We all have crap: I literally have all the bags of baggage, in all the colours, in all the sizes from my voyage thus far.

But luckily for me, I love to travel. I count myself lucky, I have the most amazing people in my life now, and people I know would miss me and people that know I would miss them more. And I am thankful to know they will always be there to pick me up, hug and kiss me and accept me for all of my flaws. No one is alone, and no one has reason to feel shameful for having experienced any type of mental un-wellness. There is help, and there are people that are more than willing to support you. We all need to talk about it.

I have a great desire to interact with other human beings by sharing my story. I want to share with people who I am, and some of what I have been through. I have hit rock bottom due to failures and new beginnings and have been crawling my way back towards my place in the sun ever since. I am giving a voice to myself finally. I have finally found the courage within to stand up for myself for no other reason than to hear myself amidst all of the noise and chaos of life. If I value and validate my own worth, I instantly also validate and bring worth to everyone in my life, and beyond. Every person I have met I am connected with, and every relationship I have is a reflection of myself. I treat others with respect, love and compassion because that is how I feel about myself, finally.

It has taken me a very long and interesting path to make it to where I am today, where I am confidently able to put my journey into words and know with great confidence that I have lived an extraordinary life. We all have a story to tell and my hope is that by me sharing mine, others will share theirs as well and we will get back to connectivity the old fashioned way--by actually knowing one another. I am not shy when I meet strangers and I know I am most likely guilty of giving too much information too quickly, but I see no point in being elusive about who I am. Sometimes people accept my candid ways. Sometimes they do not, and rebound by trying to hurt me. But at least I know now that their behavior is not my responsibility.

This is the story of one life. Peoples' names have been changed for privacy but rest assured they all exist somewhere and have all contributed to who I am today. Because one life matters, my life matters. Nothing I have done has been easy or painted with broad strokes of beige. I have lived through so much and am now thankful for the colourful experiences that led me to this moment. It has given me a terrific story to tell, and hopefully someone will enjoy reading it as much as I have enjoyed writing it.

CHAPTER ONE
I LOVE YOU

I have always felt that I loved differently than most people. I can't say I've ever feared love the way I have seen a lot of people be fearful of it. I have always loved openly. My soul and my heart have always been open. I have always offered my love freely and easily without question or hesitation. The downfall to this is that most people have been unable or unwilling to accept or even understand this, let alone reciprocate, so usually I end up being heartbroken, misguided and used. And people, generally speaking have been completely satisfied with simply taking that love and devouring it like a moist piece of chocolate cake during a food crisis.

I have always attracted, and been attracted to, the darkest, saddest, angriest people--those that felt terrible about themselves and the world around them. Subconsciously I suppose this was due to my intention to pour my love into every small fissure in and around that person in order to help them feel whole and to feel loved. I have always had so much love for others. The altruistic side of me hoped that if I could just share with someone my abundance of joy and love, that they would feel better about themselves and their world.

As it turns out, a very large majority of these people simply have not known or understood what to do with this sudden influx of something so generally "good". One of two things would happen, either push-back because of fear and a disconnection and anger directed towards me, or else complete consumption of my soul in an attempt to drain every ounce of it out of my body like a leach or a vulture feasting on the spoils of my marrow.

I have experienced all kinds of levels of love, from the tame "curious interest liking" to the less tame "obsession compulsion liking". Infatuation is not exactly as exotic as it may sound. I have found myself to be the target of someone's undying, all consuming adoration, and it is unbelievably unappealing. Usually it starts off like a romantic relationship on speed, showering me with compliments and platitudes. They want to see me all the time, and will stop at nothing to even get a whiff of my intoxicating scent, although I am not exotic. I used to get bowled over with the flattery of it all, but now I recognize how short-term it is.

I am actually just a human being and as soon as someone decides to see the truth of my many faults and kick me off the pedestal they placed me on, they turn away just as quickly as they came. And I have been left wondering why I feel so abandoned all of a sudden. I have experienced the even uglier side of infatuation that is commanded by control and fear, and it is not something I ever wish to experience again. I know the feeling of straight lust and consumption of thoughts of another, where every moment of contact was this intense, cosmic, karmic event that left me pining for more. It is like a thirst that is never truly quenched, like an itch that is never relieved, and I do not wish to be bitten again.

I give everyone the benefit of the doubt without prejudice, to a fault. I used to fall headlong into a complete mess with friends and with lovers. I made terrible choices in friendship connections, just as much as I did with romantic partners. Even up until a few months ago I was searching subconsciously for deep and lustful connections and it has been rather difficult to refocus my thinking not to be continually tripped up by the same issues. It absolutely kills me to not pursue with all of my energy those people that I am most interested in, but historically any time I have done so it has not ended up superbly fantastic. I am reminding myself to look past the plume and palette of an extravagantly well-choreographed song and dance that is the beginning of any new relationship. I am tired of hearing words with no actions, or words that are completely contradictory from actuality and reality. I am learning to wait, watch and listen to what, if anything, is being cleverly disguised between the lines.

I value others' successes, and I am proud of their accomplishments--when it has actually been accomplished. When it turns out to be a smoke and mirrors show full of empty promises and quasi goals it is rather disappointing. I don't know why people have such a problem being honest. The problem is not that they are imperfect, the problem is when someone is not willing to identify these imperfections as what makes us all unique and beautiful. I would rather know the honest everything about someone than get to know a false personhood because they were afraid they'd be judged. I love people regardless of flaws and most people are deeply flawed whether they care to admit it. Even those that have done me the most wrong I hold love for them somewhere. Well, almost all of them, anyways.

Most people tend to equate love with only romantic significance which is odd to me because I have always held love and had love for virtually every individual I have spent time with. People hold on to the phrase "I love you" for that most burning moment in a courtship as though saying those words somehow solidifies their stake in the value of that bond, just as marriage is supposed magically to keep two people together. If the actions are not behind the words, the value is lost. I have not, nor will I ever, hold my love hostage from another living thing. Love is not a tangible limited resource that has an expiration date. Love is everywhere and in everything, all the time, just as hate is. I just choose to focus on love, and I always have.

Love is not scarce, but the idea of scarcity is what drives us at our most primal level. It is part of our fundamental basic need to require a minimum level of food, water, shelter and safety. Feeling secure, however relative, is a huge force propelling some of us to always seek "more" of whatever it is we feel we are most deficient in. There have been times in my life where I have felt no love, not even a trace amount. There have been times when I have felt that I was not worthy of love, as there have also been times I have been led to believe that no one could possibly love me so that I would be too afraid to leave. I have always sought love, a love like mine, and maybe one day I will find it.

I may have believed the comments that were being fed to me about not being worthy, because maybe it was true. What had I done that would validate my being worthy of someone else's love? I now know that being myself is enough, and when I reflect on the things I have accomplished, and tried and failed miserably at, I am still a largely brilliant human being. I no longer believe the fallacies that are thrown at me, most of the time. I have had to work on changing this inner dialogue to make it very hard for anyone's words to penetrate my heart.

Everyone is deserving of love. Love in its purest form is positive, supportive, understanding, and accepting. Love is not control, love is not ownership. It is respect, honesty, kindness and virtue. A man I once dated popped into my life again recently, and has given me a gift, the gift of knowing not every relationship I had was a terrible flaming, crashing hydrogen blimp disaster. I made a very rational decision about the course of our relationship, and we handled it both with grace and dignity, and that was that. It reminded me that I am capable of dealing with the end of a relationship with maturity and respect. I have had a few like that, but they were all overshadowed by dealings with police, international divorce, and lawyers in others.

It seems as though we find ourselves in the thick of some intensely polarizing times. People are beginning to hate each other for reasons that baffle the mind. Large groups of people are becoming intolerant of everyone and everything. On the opposite end of the spectrum, because of this deep and troubling hate there have been large groups of people that are beaming straight-up love from their stomachs like they were doing the Care Bear stare. And I myself love those cuddly bears.

I still can't help but be rattled by the bigotry, narcissism, racism and hatred that seem to be consuming the world at the moment. I was up in the middle of the night for hours after the British referendum, worrying about every trickle down possible effect on every aspect of how the lives of everyone I know were about to change. There are food shortages in some countries that have the population turning on each other to stay alive, and every country has the nukes, and it's all because of hate, lack of knowledge, fear, and lies. I do identify that this is all cyclical, and it is also maddening. Change, great change comes after great unrest. Anyone that puts someone against someone else is to be met with caution. We are together in this experience, not one of us is alone, so to be fed the lies that someone does not belong to anything else does us all great disservice. I have grown weary of those that try to divide, and I understand that the only purpose it has is to empower themselves and no one else even if it is disguised as for the "greater good".

My only hope is that the choices we make in the wake of these great changes are done with the right emotions and balanced education to help us in our decision making. I can't discount compassion and I can't discount love. I honestly don't know what will become of the world, but I do know what I can do for myself within it. I can be a good person, I can continue to love and grow tolerance and respect for others. I can learn a new religion, a new language and a new culture and I can continue to support those I love and be supported by them. I can try to do and be everything that I would like the world to be, and not succumb to those who would like me to be different. I can speak and I can have a voice through written word and by talking with others. I can understand differing points of view, and not be hurt by them.

I have started to appreciate myself and have put more emphasis on paying more attention to my own needs. I consistently do things I want to do as opposed to things I feel obligated to do. I go for spa days and look after my body and my soul. I send weirdly hilarious messages to people I barely know without worrying just how strange they perceive me to be. I know I'm gently weird in an honest "sciency" kind of way, and I am good. I know my intent is never to harm. If they don't respond, I go about my day and try not to dwell on being rejected or misunderstood because I know the right people will understand. I may have censored myself dutifully in the past, but I'm not doing this anymore. I am a good person, and I would like to draw other likeminded people near me by being myself.

The only tattoo I ever had done on my body was when I was 17, and it is of a rainbow heart. It was the only thing I could think of that I wanted to permanently affix to my body. To me it represented my love for everything and everyone. I didn't even know of the pride flag back in 1997, but it just made sense to me. I have the ability to give and receive love, and so I do. And so the tattoo remains, just as I do. The colours have faded with time, as has my naive conception of the world, but still it remains--as does my love.

CHAPTER TWO
MY FIRST BOYFRIEND

I was a bit of a late bloomer. I didn't date often or much in high school. I felt myself to be pretty gawky, and very slender. Not exactly the shapely womanly figure guys tended to be attracted to in their very hormonal teenage years. I was okay with being solo as a young lady as I found it easier that way. I was very extroverted and confident on the outside, but extremely contemplative and anxious on the inside. I didn't want to be judged by someone near to my heart, so I kept most at a distance.

My first "real" boyfriend happened when I was involved with an international cultural exchange when I was 19. I had been on again-off again with a fellow from high school who was very good at non-commitment and mixed-message sending. It was infuriating as much as it was exhilarating but after some time apart it just became frustrating. On this cultural exchange there was a young French Canadian man that took quite an intense burning shine to me. It was the first time I'd known that someone was truly interested in pursuing me. At the time, I was an assistant in a high school library in a very small town in the northern Canadian tundra. A very disconnected, isolated community of farmers and not much else. He spent his days compiling and sending me love letters, which he painstakingly put together in mostly English, his second language, with sprinklings of "Spanglish" and "Spench". Our group was largely trilingual, although I often shied away from speaking out loud anything other than English.

It was an extremely raw time in my life that challenged everything I thought I knew of my frail adolescent identity, so looking like a failure in languages was not something I felt comfortable with at the time. His love letters were so incredibly full of passion I was almost overwhelmed with the power and force in which he pursued me. It only took a few months before our first kiss, on the driveway of a homestay after a night out at the only pub in town. After that, he inundated me with requiems, poems, and verses from the richest literary minds in the world. He took me to elaborate dinners, bought me small gifts, and serenaded me in broken "Frenglish". I was almost overwhelmed by the energy behind the whole affair. I had never been chased before. And it was an amazing feeling, I was almost high whenever we were together, and pined deeply when we were apart, like a fervor or being in a desert with no water. The first few months were amazing, lovely, all consuming. I had never felt a love like this before, it was foreign, and I wasn't sure how to handle it.

Once our time in the icy tundra was complete and we had made our way to our counterparts' native country of Cuba, things started to get a little more complicated. He would still dote on me and do very thoughtful kind things for me. He would spend countless nights at my side, and whatever time in the day our schedules would allow. I worked in a natural medicine production facility, which in Cuba was a 3 room house with boxes of brown jars and homemade labels. I spent my days writing out labels for tinctures and salves, antitussives and lotions. Half of the days we either had no power, or labels, or bottles, or water to make the potions with. And nothing could be done about any of these shortages. The Cubans were used to this after all. Shortages were a common mainstay. A lot of the time we would do each other's nails, and they would tell me, in Spanish, all the latest gossip in town including who was sleeping with whom (and I learned everyone was basically sleeping with everyone). They quietly told me stories about the regime, about Castro, about life in Cuba.

Externally, and in public, it was all roses. Life couldn't be better as they had so much and were so thankful. "Patria o Muerte!" they would beckon from the hilltops. But in their living rooms, they were tired, hungry and bored of a life more average than most. It was a hard place to be for me, who had experienced growing up in a developed nation, where salaries were not capped across various careers, where food choices were almost always too plentiful, where holidaying in another country or continent was pretty normal. Having a credit card, easy access to toilet paper and electricity, constant and consistent electricity to keep food cold and safe, and clean water. I can't even count the number of times I had food poisoning from eating food that had not been kept cold due to the power turning off without notice for various lengths of time and therefore no refrigeration, but I ate it anyways because there was literally nothing else. And it was all rationed. Telegrams were still the most widely used form of communication with the outside world, and I was able to experience sending quite a few in my time there. It was interesting to learn how to punctuate a telegram, and how to construct the body of information. I paid by the letter to send it across the airwaves, and it was a real treat. Telephone reception was not always clear and there was no internet anywhere that I can recall.

Every day of the week was a different foodstuff ration day, and every person would have to go down to the bread shop or the egg store with their little booklet to receive supplies in exchange for a stamp and a signature. There were certain days of the month where young men on bikes pulling massive, plastic open containers full of some kind of "soylent" type drink would come around and dole out a litre to every house. I loved it as it tasted a bit like yogurt drinks in North America, and was more palatable than any other choice. No one could really tell me what it was exactly but we all consumed it regardless.

There was one day every few weeks when everyone would get a portion of something they called hamborguesa which was supposed to be ground beef, and they all thought it was ground beef, but I knew what the texture and taste and colour of ground beef really was, and this was not it. It was more like texturized pink playdough and they loved it because it was considered a rare treat to receive. All the food had flies on it, dirt, grit or hair in it but I ate it all anyways, and somehow even gained weight for the first time in my life! My homestay mother was so proud, she had herself a little white gordita. If you were plump you were rich, and so I was a symbol of the wealth of their house. I was also drinking a fair bit more than I ever had before as well, mind you. When there was no food, it was easy to fill up on cheap mentha (a mint liqueur that was probably 5 pesos/0.25CAD for a 750 ml. bottle) or the local Cristal beer.

Christmas was quite the event in Cuba. There were no Christmas trees, or stockings, or piles of gifts. But there were gigantic social functions in which the whole community would get together to cook a lot of food, drink a lot of beer, and dance until the early hours of the morning. Christmas Morning began with me waking to two of the village men knocking impatiently at my door. It was time to prepare the evening meal which like most community gatherings involved slow roasting an entire pig. They had bestowed on me the privilege of driving a knife into this pig and they were here to take me to where our dinner was waiting, alive and walking around. A small group of us were standing in the backyard of one of the local townspeople and they had procured a very large, well fed animal. I introduced myself to the pig and began to really worry that I would not be able to go through with ending its life. The Cuban men had been chatting and explaining how it was to happen and what would happen after in terms of readying the meat for cooking. I was really only half listening when they handed me a large metal hook tipped knife and quickly grabbed the pig to tie its ankles together. I began feeling even worse for this, but kept reminding myself that I ate meat and this is where meat comes from as a way of justifying spilling this animal's blood on Christmas morning. They kept shouting instructions at me over the sound of shrills and squeals this pig was making. I began to cry and couldn't bring myself to get anywhere near the animal's rib cage, over the heart. One of my counterparts grabbed my hand that was wrapped around the shaft of the knife half-heartedly and thrust the blade through its ribs. I could feel the knife as it ripped apart muscle, tendon and bone, through the body cavity into the heart. Blood spilled forth from the pig and his legs continued to move for a few moments after.

I felt so terrible and my stomach sank and I began to cry harder knowing she was no longer breathing, walking or talking. They waited a few moments for the blood to fully drain out of the body and then they hoisted her body onto an outdoor table where they began to dissect the animal one organ at a time, all the while explaining what nutrients were available from each organ and how they would use that organ to prepare certain dishes. It became more clinical, and more of a biology class once we hit this stage, but I still felt serious remorse for having any hand in the killing. They scrubbed the skin and washed off the cavities, splaying it open along the sternum in order to lay it flat on an industrial sheet to be cooked with garlic, onions and oil. It was definitely a different way to spend Christmas than I was accustomed to. Most of the afternoon I spent doing things to distract myself after the reasonably bloody morning. I spent time with friends and family and did the hair of my host sister in preparation for the community dinner which would have my pig as the centrepiece for dinner. We arrived at a great hall around 5 pm where about 100 people had gathered to take part in festivities and music. Shortly after we arrived my pig was brought out with the aid of four men carrying a corner of this gigantic metal baking sheet. They had put an onion in her mouth, and they spoke honorably about how I had a hand in preparing the meat. I could only manage to eat a bite though, and became a vegetarian shortly after.

I now know where meat comes from, not that I didn't before but this was a new experience with how to go about obtaining meat. I understood those tightly packaged clean little barcoded parcels of goodness didn't just grow on a tree, but it was something else altogether to be the one to see, hear and feel the animal actually die. Maybe if we were more connected to our food, we would have a healthier relationship with its consumption. I don't eat a lot of red meat, or any meat much. I am not a staunch "anykindoftarian", I eat what I want mostly, when I want to. I don't eat much if any "junk". I just know I feel healthier after I consume plant based food, so that is why I do it. No ethos, no wanting to be part of any group of food eater, no trying to appease any one or any one thing except my own body.

Frenchie was still very into being with me, but it started to become unhealthy. I wasn't really allowed to do things without him around. I couldn't hang out with other friends unless he was there. The Cuban men always whistled and cat-called at me, and he became increasingly hostile at these mostly harmless actions. He started picking out little things about me and letting me know his opinion on them. How I looked, how I did my hair, the length of my skirt. He started to withdraw and ration the interaction we would have with each other. I was constantly asking what was wrong, if I had done something to anger him. And I always had, apparently. In ways he couldn't really pinpoint, or in ways I never would have comprehended would have appeared to him that I had been disrespectful somehow.

I'm fairly easy to talk to, I make friends extremely easily, and I feel comfortable chatting to anyone, anytime, anywhere--which is what he said he admired in me when we first started dating, but now all of a sudden it was a major reason for discord. He accused me of cheating on him almost on a weekly if not daily basis. He would throw massive fits of rage and jealousy any time we were out, or if there was a public function we were attending.

I remember having to leave a quincanera (15th birthday party-a very big deal) of my Cuban sister in tears after Frenchie left in a rage over something I had apparently done to offend him. And I felt doubly awful for having to leave such a special community event. Now I know it was just another mechanism of control. The party itself was a community affair. Friends and relatives came from miles in every direction. Piles of food were prepared in anticipation of hundreds of guests. Numerous pigs were cooked in honour of the occasion, and I spent days folding cardboard boxes for people to pile up with congris (a very typical dish of rice and beans), pork, fried bananas and sliced tomatoes. There was a huge sheet cake made that was as large as a dining table and it had been decorated simply yet festively. My Cuban sister spent hours preparing her makeup and outfit for the occasion. Dozens of kegs of Cristal beer were brought in. The house was hopping! An uncle brought a stereo system that rivalled any super club and everyone spent the afternoon and evening dancing well in to the night. Photographs, which were scarce in the vast majority of homes were taken to mark the occasion, kisses and hugs were given. It was truly a joyous occasion. But I had to leave, to run after my now furious for some unknown reason boyfriend. I spent the majority of the rest of the afternoon away from the party and sitting on his bed, in his homestay families house, listening to him complain and cry about why he felt my attention should have been on him, and not on anyone elses. I was just so tired, I didn't know what else to do. I felt so guilty leaving my Cuban sister. This was the one day I should have been there for her, but I allowed myself to be pulled away.

On another occasion my sister and I went on a hunt to find me peanut butter. This very average Canadian accoutrement was excessively hard to find in Cuba and we had heard through a friend of a friend, that there might be someone across town that might just make it. We set off mid-afternoon, walking along the town's roads, stopping here and there at different houses to ask very politely if they had heard of such a person that makes a kind of spread out of peanuts. Although it was February, it was still quite hot outside and the heat of the sun made my shirt stick to my back from sweat. It was late evening, around 8 pm when we finally found the house we were looking for. Not too many people had phones so interaction was direct. We walked up to the house on the left hand side of the street. There were sheets laid out on the asphalt with peanuts in the shell drying in the now faded sunlight. We knocked on the door and my Cuban sister had a swift conversation with the woman at the door, exchanged some money and passed me a small rectangular parcel wrapped in parchment paper. I opened it up and found that indeed this was crushed peanuts and a whole lot of sugar, formed in to a kind of candy bar. It was not really anything like peanut butter, but considering the five hours it took to find it, and the cross town adventure we had, it could not have tasted better.

The rest of the time spent in Cuba was the same day in and day out. We would have work or no work depending on supplies, a short visit to the Cayos on the way out of town and a few days in Havana, eating mini pizzas from the street food vendors and purchasing every possible souvenir with Che Guevara's face printed on it. It should have been a simple existence, if not for the man drama in my life.

One thing has to be said about my time spent in this small island nation that much of the world has little to do with, was the impact it left on me. Nowhere I had been until then, and certainly nowhere I have been since came close to having the same sense of spirituality as Cuba. The people had absolutely nothing. The few things they did have they used and reused, and repurposed when their first purpose had no use. They loved so hard and despite having next to nothing, I truly felt and believed as though they had everything. They were calm, collected, and full of laughter. They did not worry so much about the things most people in the Western world take heavily for granted. They did not war with each other, because there was a deep sense of belonging, of purpose, of community and of spirituality-and to no God in particular. Yes, there were the common denominations of churches and such but it went deeper than organized religion. This was a passion and a love that went further than appearances and name brand labels. My host family loved me, cared for me and made sure they did everything they could to provide what they could. And they did an amazing job. I sobbed relentlessly when we had to leave. It had been such an emotional roller coaster, and now it was time for us to go. I felt ill prepared to go back to Canada, although the whole time I was in Cuba I pined for things back home.

On the flight back to Canada, I ended up sitting between two burly rig workers from off the coast of Varadero. They were funny, delighted in flirting with me, and we had some great conversations. Frenchie was livid. He started scrunching up notes written on bits of paper and throwing them at me. He called me a whore amongst other French slang for speaking with these two guys. He first ignored me for days, threw things at me, and then yelled at me, just for talking to two men. I was beside myself. I felt dirty, shameful, not worthy of love or affection from anyone. I didn't understand why I had been so vehemently outcast and publicly smote from the first person I ever thought cared about me, deeper than any other before. I begged and pleaded for forgiveness. He wouldn't even look me in the eyes when we spoke. He disregarded everything I had said. My feelings did not matter, as I was the one that caused him harm.

I felt myself suffocating, thinking he was the only human on the planet that could ever give me the air my lungs so badly needed to survive. He loved me so much, why would I do anything to cause him so much hurt?! I am a terrible human being I would think to myself. I apologised for it seemed like forever until he would eventually forgive me. Back in Canada we couldn't stand being apart, we had to be together, because as he would tell me, his life had no meaning without me. I gave him purpose, and joy, and everything he ever wanted or needed in life.

So I moved to Montreal. It took two months for us to find an apartment, and about the same length for me to find a job as nanny for a family. We shared a bank account, we shared an apartment, and we shared everything we could together. I had friends, but I rarely would see them on my own. I was cautious and did not want him to think that I was out doing things that would hurt him in some way. I was walking on eggshells constantly. I did not want to feel his cruel and cold wrath again. But I did, over and over again, and for anything. Liking the "wrong music", wearing the "wrong pants". He cut my hair because he said it would make me more beautiful. He dyed my hair because my natural colour was "kind of dull and mousey". He spent thousands of dollars on electronic equipment and hydroponics to pursue his own interests-such as photography and marijuana growing in our apartment, but had no money to spend on groceries or hydro bills.

I started withdrawing more and more. He started talking irrationally about things that I have no idea what they even were meant to be about. He would stare off into space with a glazed eye, he started quoting Nietzsche daily. The Insufferable Lightness of Being was his favorite work by that author. We no longer went out, and he started talking poorly about his friends, people that had known him his whole life. We never went to restaurants, and I wasn't allowed to buy clothes. He bought his sister hundreds of dollars of gifts and technology, but wouldn't give me a card on my birthday, not even a hand drawn home made one.

In the summer of 2001 Montreal had massive rainstorms. The thunder shook buildings, the lightning lit up the whole town. My brother and a good friend of mine from school came east to visit me, and for a few days I was having fun. During one of the storms we took to the streets which were flooded in thigh-deep water flowing down in every direction, stopping traffic. I felt free. I had lost weight though, and they had both noticed. I cried easily, for no reason. But I was so afraid of admitting I needed help getting myself out of this situation. I was so trapped in my own ego of my choices and my predicament.

I had started looking for jobs abroad, and Frenchie had installed tracking software on our computer and confronted me once he had spied on my internet activity. It was a massive blow out. He screamed and yelled and accused me of cheating on him, of possibly abandoning him, of leaving him to die. He stormed out of the house and I was so stricken with grief I huddled in to the fetal position and cried for an entire 24 hours. He returned the next day, wild eyed and disconnected. His face went sideways as he sat on the edge of the bed and he smiled at me, told me how he tried to love me, only it wasn't enough because I was a dirty whore that was not worthy of his time or his love. I had no worth, and because I wanted to break up with him, I had to pay the price.

He took two bottles of pills out of his pocket, and started putting them one by one in his mouth, and swallowing them, smiling at me all the while. I was stunned. I didn't know quite what to do, so I called 911 and told them what he was doing. They asked me if he was trying to end his life. My hands and whole body were shaking, so I asked him out loud on the phone if he was intending to end his life. He smiled, nodded his head and continued to swallow pills, telling me it was my fault he was doing this. I was making him kill himself, because he loved me so much and I was weak because I couldn't love him the way he needed to be loved. He grabbed the phone and threw it against the wall. Then he grabbed me and threw me across the room and into the wall, and ran out in to another rainstorm. The paramedics showed up and I was actually very calm in that moment. Like the eye of a hurricane. I told them he had left, and there was not much to be said. I didn't bother reporting him assaulting me because at that point I was in such shock I wouldn't have known what to say anyway. And I still felt like his pain and suffering were more important than what he had done to me.

He called his parents about eight hours later, from one of the small islands off Montreal. They picked him up and had him admitted to the psychiatric unit of the hospital. He called me on a daily basis, letting me know that this was all a big mistake and that he accepted my apology for making him do such a thing. We were going to be together forever. We were going to have babies, and a white picket fence, a dog, summers in the Laurentides. It was the twilight zone. I stayed with my very gracious employers until it became unbearable and he had been released from hospital. He kept threatening to track me down and he sat outside their house. This served as a huge embarrassment more than anything else. I felt so ashamed to have dragged this family in to my mess, so I went back west.

He continued to call and to send letters. He returned my belongings COD on the Greyhound, only he had cut up everything before he sent it, and filled the boxes full of rocks. So I spent hundreds of dollars on worthless boxes of garbage. He even managed to completely dismantle the frame of my bicycle before sending it back, worthless and un-rideable, but I still had to pay for the shipping nonetheless. He would harass me on the phone, and say he was going to show up on my parents' doorstep, which I was also ashamed about, so I applied and was offered a job to nanny in Rome, Italy which I took. I had failed miserably and was sure that this was the most terribly distressing, soul-shattering experience I would ever live through. All these years later I still hear from him occasionally. He is married now with a child and another on the way.

The legacy of this experience is that it set up a pattern for me, a pattern I am now concentrating on consciously breaking. I was able to walk away with little attachment to Frenchie, but it took me years to get over the pain of the whole experience. I felt so useless to help him and I wanted so badly to be able to help him, to take away his pain and his suffering with my love. However, I lacked the skills and he was not capable of even asking for help at this point. I grieved and went out a lot trying to escape the pain of thinking I had caused someone to want to try and end his life. My friends tried very hard to help divert my attention to other causes, but with little success.

A couple of weeks after I returned, one of my dearest male friends from elementary school shared a very tender moment with me while we were walking through the park on a warm summer evening. He had let me know he had made the decision to let the world know his love was with men and not women. He shared with me how painful that entire process was for him and what it was like to come to that kind of understanding. It is never easy to face truths, no matter how much of a part of you they always were. I felt absolutely honored to be someone he felt he could share that openness with. I spoke of my aching, and he spoke of his newly acquired freedom to be. We connected because of the challenges we had faced and we felt comfort in knowing each other's pain and struggle. I once dated this gorgeous man for a few short days when we were still in our youth and I still to this day love him more than I could possibly express. Although he has had struggles he still remains to be one of the most vibrantly positive people I have ever had the luxury of having in my life. I often joke I will marry his cat because he doesn't seem to want a wife and child and it always makes us both laugh. The cat is pretty cute, to give it some credit.

CHAPTER THREE
SEPTEMBER ELEVENTH

On the evening of September 10, 2001, I was spending some time with my most favorite family in Montreal. I had flown there from out west, on my way to Europe to begin a new life, one that was safe from a mentally unstable ex-boyfriend, one full of hope and promise. I was going to learn Italian, and become a native of Rome. Eat gorgeous food and wash it down with even more gorgeous wine. I was going to be able to help guide three children through school, help them improve their English. It was going to be great!

I was so full of enthusiasm, and trying my hardest to shake off the trauma I had just gone through. My Montreal family had eaten dinner with me and dropped me off at the airport, with tearful hugs and well wishes. I was so truly thankful for their support. The flight was really not much to speak about, there was a bit of turbulence, but nothing major. It was the red eye, but I couldn't sleep. I was thinking of all the wonderful things I was going to experience, a life reborn in a European metropolis. I had taken a dimenhydrinate to help me sleep, but instead it just made me wired and I didn't catch a hint of sleep for the duration of my travels.

The plane landed early in the morning local time. The line for immigration was totally packed. I thought it was odd, but as this was my first time in a European airport, I supposed it was just the way it was. The family I was to be working for did not show up at the airport to collect me so I hailed a cab and showed the dark haired, slender driver the address where I was supposed to be going. We weaved through the cobblestone streets of downtown Rome, past the Colosseum, and past open air restaurants and cafes. I was exhausted, but I was excited. The air did not smell fresh, but it was warm and definitely cosmopolitan.

And then we were out of the city centre. Buildings became more sparse, countryside more abundant. For a moment I did have the thought that this taxi driver was leading me astray for some misguided reason yet still, we drove on. It took an hour and a half, or more like an hour and 37 minutes to reach a gate, at the end of a road. The taxi driver leaned over the back of the seat and asked very puzzled if I was sure this is where I wanted to go. All I could do was shrug and point to the address I had printed off from the email the family had sent to me a few weeks earlier. He pressed a bunch of buttons on an intercom and the gate slowly swung open. We passed about a kilometer of immaculately groomed hedges, shrubs, trees and lemon groves. There was a team of gardeners on their knees pruning this or that. I noticed as the long and winding driveway came to the end that there was a castle to my left. Where was I? Did everyone have a castle here? It was Italy after all. The taxi driver got out and spoke to one of the gardeners, a leather-skinned man. He pointed to a house up and to the right of where we had stopped. The taxi driver got back in the car and continued on to a driveway. There was an open sided parking garage to the left, with seven brand new Mercedes Benz parked in the garage. I gave the driver about 200 lira for the fare, and he was on his way. I stumbled out with my purple backpack and found the door. A maid met me and scuttled me inside. She was Albanian, I would later find out, and very young.

The mother of the family welcomed me with grand gestures and spoke regally about the property, the grounds, the family and the country I now found myself in. I was so tired and overwhelmed I really had no idea what was going on. They hadn't yet prepared anywhere for me to live, let alone a bed to sleep on. I can't even recall what time this was now, sometime after lunch presumably. I remember sitting down on this empty uncovered mattress that was in a space that was to be my room. I looked around and noticed everything in the house was made of marble. Everything. The floors, the counters, the stairs. There were chandeliers in every room, tapestries hanging from every wall. Large antiquated pieces of art and armour and everything you could think one household must have. There were lemon trees on all of the massive concrete pillared terraces that led out to sweeping views of golf courses, helicopter pads (yes, more than one helicopter pad), tennis courts and lavish green gardens.

I hadn't eaten in a very long time, and hadn't slept in probably two days due to excitement and crying babies on airplanes. I was sitting on the edge of the mattress as the team of household help was trying to put together a room for me, at that very moment. The sliding door to my room that led out to a tennis court and a garden was open to let in fresh air and I heard a young girl's voice shrieking loudly as she came bursting in. She was a beautiful curly haired blonde, olive skinned, blue eyed young girl who was the daughter of Miss Universe, because obviously Miss Universe was their neighbor. She was speaking so fast and in Italian that I had no idea what was going on, so the help plugged in a television and turned it to CNN Italia.

The second plane was going in to the second tower at the World Trade Centre. And in that exact moment my jaw dropped, and every single red blood cell ran away from my body. The room was silent for about two minutes. No one said a word, in any language. And then they went on like nothing had happened. The help continued to dust and clean windows. They wouldn't even look at the television as I sat there in complete horror. My only thought was I am in a foreign country, this is World War III, and I have no way of doing anything. I had friends in New York, friends that worked not too far from that scene. I was just full of dread. I was afraid to cry though. These people had known me for 20 minutes and now I was falling apart. They shrugged it off: "it's so far away" they'd tell me. "Nothing to be done" they would say.

I called my parents to let them know it was not my plane that had been over the Atlantic Ocean earlier that had met its' untimely fate. They were relieved, but I was not. I thought if the USA was being attacked, who would they then turn around and attack? Italy? Not likely, but in times of such great uncertainty, nothing is clear.

The family was not terribly nice. The children did not speak more than a handful of English, and I spoke even less Italian. But I tried to assimilate quickly. It didn't work, after three days and a bunch of arguing with a culture that valued smiling above all else, even when they were calling you names and telling you how useless you were, after everything I had been through. I'd had enough. I had a conversation with the young Armenian maid in the cover of night and we spoke, and somehow communicated about the realities of this family, she had basically been sold to them, but considering where she came from she was grateful for the job. The next morning I had already packed my bags and told them to please have one of their drivers take me to a hostel or somewhere as I was done. They were arrogant and rude and I was not staying. They only agreed to take me to a taxi stand at the edge of town, and that was just fine with me. My dad, with all of his foresight and planning had given me a guidebook to Rome. I got in the cab, opened a random page, pointed to a hostel and gave a thumbs-up to the Taxi driver.

I ended up at the Yellow Hostel in downtown Rome. It had a very narrow old elevator, the kind you had to pull the metal grate across manually to operate. It was very cool. I had probably $500 CAD to my name, and a credit card with that much left as a limit. I couldn't fly back to North America just yet, due to the whole terrorism thing that was going on. I had no plan, no contacts, no language skills, and no preconceived notion of a plan B. But here I was, in Rome.

I made friends with everyone at the hostel pretty easily. One was on her way to Southeast Asia within a few days and looking to offload her Shoestring Guide to Europe by Lonely Planet, for 35 lira. So I took it, and decided I was going to make the best of it somehow. Because I had no idea what I was doing, or where I was, or even if I had packed the right things to be now all of a sudden backpacking across Europe. I just started talking to everyone, asking where they had been, what they had seen, what they wanted to see.

I met a lovely young man from Reading, UK who was on his last days in Rome and we decided that tomorrow we would go by foot around all of Rome and see all of the sights. For 18 hours we walked the city streets. We literally went to every single touristy destination you could go to. We skived in on tours of the ruins of the old city, we went into every major gallery and exhibit. We dined at every recommended cafe and restaurant and talked openly about everything all day long. We even managed to find ourselves in the midst of a protest walking down the cobblestone street that led to the Colosseum. We ended up cuddled on a hill of grass at 11 at night. The Colosseum was lit up fantastically. It was a beautiful moment. It felt so nice to have discourse with someone, to feel connected simply from spending time together. Both of us knew we would probably never see each other again, and we just hung in that moment together, taking in the lights of the Colosseum. He left the next morning for home and I decided to head south to see Naples, the Amalfi Coast and beyond.

CHAPTER FOUR
EUROPE

My time spent on the other side of the pond was definitely the most happy, carefree, explorative, free, clumsy time in my life. I had gotten there almost as though by mistake, except I know it was completely intentional. I was free in Europe, very few people knew I was there and I knew no one. I guess some would see that as being a scary prospect but I didn't. After my first experience with stalking and abuse it felt absolutely amazing to walk down a cobblestone street in Rome without having to worry that someone who knew me was right behind me, taking note of what route I had taken or where I ate that day.

Rome was architecturally stunning. Despite knowing nothing of the city, having no plan, and having no money, I had no problem feeling comfortable as a tourist. I bought pasta from the supermarket for next to nothing, and since I was a vegetarian meals were generally pretty cheap. I wasn't terribly into fancy coffee and wasn't really going out on the town all that much as I hadn't really connected with anyone other than the young man from Reading I had spent that one day with when I had first stumbled in.

Eventually I started catching trains to places. The first place I thought would be worthwhile visiting was Pompeii. I didn't really know what to expect and was severely underprepared but found some other young people on the train on the way and we wandered around the ruins together. One hardly spoke English, but he shared with me his pears and apples that he had brought in his day pack. I had brought my entire backpack and had left a whole bunch of luggage in storage in Rome at the hostel, so I had to squeeze all my worldly possessions into a tiny half locker at the ticket booth near the entrance to Pompeii at the insistence of the admissions staff, possibly for fear I would just set up shop somewhere and become a resident. The ruins are totally creepy in the most exhilarating way. Everything is literally frozen in time. Bodies encrusted in lava are held in the same positions as when they met their fate as a wall of hot magma engulfed them as well as their houses, irrigation systems, pots, water containers, brothels. All the original artwork was intact as were the arenas that are now completely empty and the doorways that were all apparently for short people. I felt like at 5'8 I would have seemed like some kind of mammoth. It was still hot, being mid-September, and because it was not peak season the streets of Pompeii were not terribly overrun, but you could still buy all kinds of tacky merchandise. I never bought any of that though, not in my budget of zero dollars.

I headed to Naples next; being the home of Pizza, who could resist! The great thing I found about all of the Mediterranean is that anywhere I got off any mode of transportation there were literally dozens of people looking to rent out their penciones for anything reasonable that you'd offer them. Naples was no different and meeting people was just as easy. Most places I stayed also offered a home cooked breakfast by the owner of the flat that you found yourself staying in, and it was at these breakfast times that I would always meet someone going somewhere that didn't mind me coming along. Come to think of it, perhaps it was strange for them to be seeing this barely 100 pound blonde white girl carrying a 50 pound backpack, completely lost wandering through Europe with essentially no idea where she was going.

Naples was an idyllic slice of Italian countryside mixed with a small city feel and I ate the freshest, cleanest-seeming food so far. The pizza was gorgeous. I had a very fresh savory pie topped with equally fresh mushrooms and I remember trying to find something without meat which was always met with curious looks. This was a long time ago, and I don't know how popular vegetarianism was then as compared to now. Everything was made with the freshest tomato sauce as though they must have actually just killed those tomatoes that afternoon. Food in Italy I found to be much simpler than anything I had been used to until this point. There were no crazy spices, they did not use a lot of salt. I had heard about the food in Italy growing up but expected it to be quite different somehow. In retrospect I understand now how the freshest of ingredients far outweigh any prepackaged, overly processed meal I would have craved in my younger years.

I found my way to the Amalfi coast via a train, a winding bus trip and a long walk around a gorgeous seascape promenade at dusk with a young couple that may have been American. They were in their mid-20s and basically hated each other. I'm not sure exactly why, or if it was just because of the being constantly together during their travels, but they did try hard to come across as though they were the happiest boyfriend and girlfriend, but they definitely didn't seem so contented. They hardly looked each other in the face or the eyes, held hands or spoke much to each other. They led me to a pencione in a small town carved out of a hole in the rock on the seaside. They had their own room and I rented a small single room that was basically a straw single bed mattress that lay on the ground and a dresser with a white and blue pitcher for water sitting on the surface. The walls were made of whitewashed cement, the ground was cold and made of finely polished concrete.

It was dark by the time I washed away the day of travel in the sink and got myself dressed. The whole town was lit in small strings of lights hung along archways and tea lights set on top of whitewashed cement stairs. All the linens on tables and furniture were actual linen cloth and there were no plastic cups anywhere to be found. Groups of diners ate al fresco for every meal on tiny terraces, some overlooking the sharply cut coastline. There were small two-person fishing boats with colorful striped paint jobs dotted all along the small pebble sea shore, and the moon illuminated the sky and reflected off the calm sea surface. It was beautiful like a painting in a gallery, hung on the wall of a Roman museum.

The next day I walked down to the seashore and was the only person on the beach. I was excited as this was the first place my body felt the Mediterranean Sea. It was much cooler than I expected it to be, and not quite as clear as I'd envisioned it, so my dip wasn't for an excessive period of time. I was wearing my most favorite bathing suit of all times still to this day, a black bikini with palm trees and hula girls on it. I honestly would give the left nut I don't have to have another bathing suit with that pattern in those colours again. After my inaugural Mediterranean baptism I walked back up the steep cliffside back to town. I relaxed in the open air courtyards and watched the villagers and tourists exchange dialogue. I spent a few days eating fresh fish, and produce from small cafes. I visited the internet cafe, which literally had the only computer in town in it, and it was expensive at about a dollar a minute. The computer was one of those blue Apple desktop computers and had 28,800 dial-up with the screeching connection sounds. I mostly just checked my email and wrote my parents occasionally to let them know I wasn't dead. The happily unhappy couple had stayed a while longer here but I moved on down the road to Sorrento, Salerno, and anywhere in between.

I caught the train to Bari as I decided to travel further south to Greece. I had heard good things, and I was excited to feel warmer still. My first time on a Mediterranean ferry was really quite interesting. BBC world news was playing on all the televisions as there was a war going on and hours of video loops of shelling and bombings in some other foreign country. I didn't pay much attention as I was already riddled with worry we were headed into World War III back in North America. I didn't really like news while I was in a foreign country, cut off from everything. I was still reeling from Frenchie, and from the Italian nanny experience. I was happier to focus on where I was going next, where I would stay and what I was going to do.

The boat landed in Piraeus early in the morning. Sleeping on the ferry with a non-sleeper ticket was not fun as it was very loud and the air reeked of diesel. There were still a few tourists despite being the shoulder season, but a lot of local Italians were going for short sojourns as well. I was pretty tired, and pretty bloody overwhelmed on the streets of this crazy port of a city. Homeless people were everywhere as were the pigeons so I opened the shoestring guide and just started walking. I did not speak and could not decipher the Greek alphabet when I first got there, but luckily mathematics look the same in any language, so finding a bus number to a hostel was fairly uncomplicated.

I stayed in downtown Athens for the first few days, chatting with people, figuring out what some highlights were, which were islands that sounded like what I was looking for. Mykonos I decided was definitely out for the time being. I was not yet in an all-out partying mode, nor did I really feel comfortable to do such a thing solo. I spent days wandering the underground Metro they were scrambling to complete for the upcoming Olympic Games. I sat in parks and read books, went to museums and monuments. I went up to the Acropolis and watched what I thought would be a gorgeous sunset over a densely populated, almost all completely whitewashed, sprawling, urban network.

The air in Athens at that time was terribly polluted. It was so thick with haze that by the time the sun hit a 45 degree angle on the horizon, it had no defined shape anymore and just became a blurry horizontal strip of red and orange. On the way down from the astoundingly large, yet small Acropolis, I saw a pack of dogs eat another live dog. Or rather, watched them begin to rip the flesh from this dog`s body, and then a bunch of tourists and myself started throwing sticks and rocks at the pack of dogs to at least give the poor other dog a head start to get away. But unfortunately I'm sure that scenario did not end well. `

I had met a group of young people while I was up on the rock, two Kiwi (New Zealand) girls and two American boys. They had met each other about a day earlier and they were all flirting with each other, to the point it actually became irritating and I chose to make other plans after a few hours. They were a little too carefree for my then very burdened headspace. I decided to travel to Naxos, Paros, and eventually spent a few weeks on Ios.

Ios is a small island in the Cyclades, and damn is it beautiful. I stayed at the Far Out Beach Club, to the south of a long reaching crescent bay facing west. They had ridiculously cheap cabanas on a gently sloping hillside with outrageously pristine views of the Mediterranean Sea. I met some amazing people, went to some unbelievably delicious seaside restaurants and ate all the feta and olives and fresh goat yogurt with local honey and almonds I could get my hands on. I was approached to work at a bar up in "Oia" (or main town) called Flames bar. It was managed by a Swedish couple in their early 30's and owned by a short, stout Greek fellow. They basically paid me to mingle with men and get them to buy me drinks. I also wore their tank top at the poolside back at the Beach Club, as this is how advertising was done. You would see all these young ladies in shirts for all the different bars and clubs in town on any given island and it was a pretty fun gig.

I made friends with a very lovely Scotsman one night, whom I am still in touch with to this day. He treated me like I was his little sister, and I loved the friendly companionship on an otherwise solitary journey. He was selling jewelry at the time, and gave me a belly button bar that had a blue crystal in the centre of a starfish dangly charm on it. So obviously I decided to go get my belly button pierced. I had actually heard there was a veterinarian in town who often did body piercing for tourists for 20 drachma cash, and he used a localized anaesthetic which made it totally painless. I sat back on the examination table in his very clean office. He explained in broken English what he was going to do before he did it. I watched carefully as he inserted the needle through and around my skin and noticed how gently he guided his instruments. It was all done in about 45 seconds, and I didn't feel a thing. I guess animal health care wasn't a huge money maker on an island with a population of 1500. Goats were plentiful, as were cats, but I hadn't seen much else.

One night, I had ended up making out with a British bloke with brown hair that I had met at the bar. He was quite tall and average looking and we ended up kissing on the beach until dawn. Dawn in the Mediterranean is the most gorgeous sight I had ever seen. The sea is spectacularly blue, and clear down to the farthest reaches of eternity. The sky is even bluer, and the sand is the purest tan colour. When the sun crests above the horizon of the olive groves on the hills, it is pure radiant magic.

I had absolutely zero interest in having any other physicality with this gentleman so we went to sleep in separate beds inside my small cabana overlooking the ocean and I snuck out shortly after he had fallen asleep to meet up with my Scottish brother in town, hiding behind a bus after breakfast when we saw him getting off another bus in town. I had crept out of my cabana without saying so much as a goodbye, and left him snoring peacefully. I wanted to desperately avoid an awkward meeting so the Scotsman and I quickly scuttled onto the next ferry over to Santorini, where we shared a suite in the most platonic way, at the most beautiful little place with its own pool in the courtyard. The courtyard was long and narrow and hugged around the long and skinny pool. Around the pool were lattice fences strewn with pink flowers and bright green vines. One of the days we were hanging out poolside, I ended up slipping somehow over the artfully tiled ledge and fell in slow motion directly into the water. Just as I felt my warmed by the sun hair wet and cold with water, the Scotsman jumped right in after me, not even skipping a beat to rescue me. I felt his arms around me as we made from a scene from a movie, hoisting me above the water line while making grand statements about my perilous condition. He still hounds me about my accidental swim to this day, although I wasn't really in need of much rescue it delights me to reminisce of the time we spent together.

Shortly after the pool incident I had taken somewhat ill, and had decided that he should leave me and continue on his journey. He gave me the book White Teeth by Zadie Smith, which proved to offer me some cheap and easy entertainment for the numerous bus, train and ferry rides I had ahead of me. He dropped me off with grand gestures at Hostel Anna along Perissa beach. He carried me in his arms like a wounded dove inside the hostel and placed me carefully on to my new lower mattress on a bunk bed. He left me laying there, thinking my thoughts about why my stomach seemed so upset.

I had napped for about an hour after he left when a most gorgeous, soft spoken, silky haired, almond skinned, gently curvy as a goddess woman tapped me on the bottom and asked if I was alright, in the loveliest and distinguished high British accent I had ever heard. I turned around on my mattress to meet Elizabeth. We chatted in the dorm room for a while about where we were from, what we were doing, and what had brought us here until I decided she was my new best friend and we went off to spent the day together on the black pebble beach Perissa is known so well for. The small dark stones were fine, perfectly rounded and so heavy. My feet and body had to work hard to navigate the small pebbles in flip flops as the weight of them would cause me to sink and slide further down in to the ocean than I was meaning to.

It was customary to pay to rent a beach chair throughout Greece and because my zero dollar budget that had been recently infused by Flames Bar cash didn't quite allow for such extravagances, I laid out a sarong my brother had given me from Indonesia that was covered in a beautiful yellow, orange and red tribal motif above the rocks. My beautiful British new best friend Elizabeth spent the afternoon telling me about her life. She was a musician from Bristol and was in the midst of a strop with her boyfriend whom she had been texting and calling on her blue Nokia 3150 while we chatted, and I found it fascinating at that time that she could be having a conversation via text from two different countries with such ease. Texting wasn't widely used as the preferred method of communication yet back in Canada, especially at such an instantaneous rate as this interchange. He kept telling her he was going to come out to Greece and meet her and make relations good again but he never did which suited her just fine. She wasn't thrilled at the prospect that he was trying to meet her as she had gone on this holiday to get some space and perspective from the relationship with him.

We sauntered back to the hostel as the sun was beginning to go down and started chatting to a superhumanly good-looking blonde male named Chris, who was sitting on the patio of the hostel. He noticed us approaching and decided to strike up a conversation as he himself had just arrived to stay at the Inn. He was also from Bristol, and just happened to know a mutual friend of Elizabeth's that had been free camping right along the ocean on another island. We made plans to leave within a few days and I was excited to have met a few people that took me in with open arms. They were generous in their attitude towards me, they were kind, they had senses of humour, and they were also open to all sorts of adventures.

I was very excited to have connected with them. We went together to eat the evening meal at a restaurant that had been suggested to us by a small group of locals we had met at a café near our hostel. We walked a few blocks through the small seaside town to a local restaurant that served everything fresh and plentiful. All the food was prepared and set out on display for us to choose from. We sipped on wine and sat in the warm open air at a large wooden table with white and blue linens and napkins. In this company I felt safe, secure, happy and relaxed. I could feel my body begin to let go, if only just a little bit.

CHAPTER FIVE
THE GRECO ROMANS

The warm wind of the sea gently caressed my cheek as I stood on the deck of a Mediterranean ferry on my way from Santorini to the much smaller island of Amorgos with Elizabeth and Chris to meet their other mutual friend. Neither of them had known the other prior to the hostel, neither of them had known that we would all be travelling together to meet Marc. The water in the Mediterranean Sea is the bluest of blue, and clear for miles down to the bottom during the daytime when the sun is at its highest. The sun seemed so much hotter and direct in the Cyclades, the air was cleaner, fresher and sweeter. The calm and stillness I felt as I travelled in those islands was incomparable to anywhere else in the world. Everything seemed tranquil.

The boat docked and we made our way by foot to Hora and then followed a sandy dirt path down past a local tavern to a very remote beach that was covered in fine beige sand, small pebbles and the most stunning view of the ocean. There were islands off in the distance and hardly a cloud in the sky. We met Marc along the beach under a canopied tarp that had been strung between trees that served as shelter. My new friend was a tall, lanky and lithe young man with an infectious grin, complete with dimples as deep as canyons. His skin was a deep olive, kissed sweetly by the sun after a few weeks of holiday. He moved fluidly towards us as he hugged every one of us, surprised to see Elizabeth and to have also been introduced to me, a random traveller they had collected along the journey. It was a joyful reunion for the Brits, and an exciting time for me as all four of us tended to get along very easily, had things in common and similar senses of humour. It was a comforting feeling. There were maybe a half dozen other people along that stretch of beach, all rather friendly, laid back and easy to speak with, one of whom also worked in the music industry in Britain and was an older fellow, reliving his globetrotting youth.

We spent much of every day sitting and laying on the sand talking. We spent the afternoons bathing mostly topless if not completely without a swimsuit in the blazing hot afternoon sun. Despite how intense the sun felt during the day it never felt too hot or too cold as we were never more than four steps to the ocean. The evenings were always balmy and the breeze was a warm 26 degrees. We ate at the local Tavern we had walked past on the way to the beach upon arrival as it turned out it was the only source of food or plumbing for miles. Most meals were completed with a shot of complimentary Ouzo or Raki from the owners which is a very standard type of licorice alcohol in Greece. The meals usually transitioned to a very peaceful nap under the shade of a tree with the sound of the sea lapping up lazily against the shore. We bathed in the sea, we swam in the sea, and we recreated in the sea. I spent hours of every day diving deep down to feel the water slowly and softly move against my skin like the finest silk. Tiny fish would dance around me and the sun's rays slid through the liquid with ease. I suspended myself in various positions under the water, eyes wide open to take in the simplistic beauty of my clean and bountiful surroundings. It was completely beautiful and I felt free.

After a few days of lazing in the sun, we rented scooters and explored up to the northern tip of the island to discover other decent locally-suggested spots. We made it a custom to stop at hilltop villages to ask the locals for their favorite destinations. We sat and sipped coffee and ate spanakopita and fresh olives while they drew out instructions on napkins. Every restaurant played endless loops of the dreamiest, blissed out ambient chill music that only helped to bring ease and comfort along our journey. Classic white windmills dotted the seaside wherever we went, and pink flowers cascaded down from crawling vines and covered the exteriors of most whitewashed buildings. Every path through every town was constructed of intricately placed, very aged cobblestone and most corridors in villages were for foot traffic only. We did not see many cars on the islands as there wasn't a huge need for them, and everyone had time for and delighted in walking. Old women wearing head to toe black could be seen sitting outside their front doors simply admiring the day, and small cats would lazily saunter through alleyways, and under their black ankle length skirts.

Agios Pavlos is a beach whose vision will be forever burned in to my memory as being the most beautiful beach we visited while on this island. The rocks drifted from tides and formed a spiral-shaped jetty that extended from the furthest-reaching point curling mathematically perfectly like the inside of a seashell. Looking down at the beach from where we parked the scooters, I was breathless from the beauty of such a sight. We spent the day lying on the hot rocks, and swam the day away. The water here was warmer than a bath and I felt completely at home immersed in it. The longer we stayed, the more I swam and the harder it became to remove myself from it.

We continued scooting from town to town, checking out more shops and alley ways, cafes and restaurants. We camped along the shore of a small beach that had a monastery built into the side of the rock along one of the cliffs. The exterior walls of the place of worship were bright white and were the only thing that stood out about the large yet humble structure. We stopped in a small harbour town just as a couple of fishermen were dragging up the boat launch a very large sea creature that looked like a shark as big as myself. I stood and looked at this creature for a time. It was laying there lifeless, blood trickling back down the boat ramp into the water, to become someone's dinner that night.

Most of the food served in restaurants throughout the islands was absurdly fresh, in fact one of the things I enjoyed so much about the Greek islands was how unbelievably fresh all the food was. Literally caught that day served with everything absolutely drenched in olive oil, lemon and fresh herbs, my mouth waters just thinking about it. I was still largely vegetarian at this time, although I did start eating little bites of fish, and the occasional slice of fresh octopus. I didn't find many options for my rather strict diet so I would have to make do and ended up consuming a lot of potato products, such as roasted potatoes in inches-thick olive oil. Most shopkeepers found it really odd that I'd never want meat in my gyros, but I'd always substitute French fries instead, topped with a mountain of tzatziki. It became one of my staple meals and I would frequent the same gyros fast food shops in the market squares and they knew what to prepare for me every time they saw me coming.

After a few days on scooters drifting around Amorgos, we took another small ferry over to another even smaller island named Anafi which had even less people and less things going on. Everything was free and open back then which was helpful in finding free accommodation along most beaches. I had left my sleeping bag back in Rome due to its weight, so spent most nights using my friends' tent, un-pitched as a sleeping bag. No one really needed much in the way of insulation when the nights would only ever drop to 24, although the temperature did seem much colder than that due to how blissfully warm the days were.

When we had arrived on Anafi, we had asked how we would get back off this island considering it was no longer peak travel time for any boats or ships. We hadn't thought about the logistics of our island hopping adventure, we simply trusted it would all work out somehow. The townspeople had suggested hitching a ride on a post boat in a few days, and we had agreed knowing full well this boat would only be docked for about 20 minutes, and would certainly not be waiting around for a bunch of random foreigners to board it.

The harbour town on Anafi was very small and hugged the shoreline very closely. We walked to the right out of town and followed a trail that led us to a beach where we spent a few days, doing all but nothing in a very blissful manner. Eat, sleep, swim, repeat. I cannot recall another time or place in which time seemed to both stand still and tick forward by simply doing nothing but enjoying the scenery. There was hardly another soul around, and we didn't mind at all. On the day we were to rendezvous with the postal boat we packed up and started walking back to the harbour. When we had gotten about half way along the long and winding dirt path that followed the shore up a steep cliff when we realized we had misjudged our travel time and had to begin running in the midday sun, I with my now 40 lb purple backpack sweating and grunting, anxiety rising. I was wet with sweat and my legs were burning from strain but we did intercept the boat just in time.

The boat was about 40 feet long with a traditional wooden keel. The sailor was a young man seemingly only a few years older than me. He had broad shoulders, dark wavy hair and a very well defined upper body. There was also a postal worker on board who didn't offer much in the way of conversation. Their actual job was to sail around the Mediterranean delivering mail every day. I doubt that it paid much but it surely would make for an amazing way to spend your days. We hadn't been the only foreigners they'd ever allowed to come aboard, and were happy to have us. If we had not run as though our bums were on fire along that dirt trail, we would have had to stay another week (which really wouldn't have been so terrible) before returning to Santorini.

We stayed at a small pencione in a small village with two double beds for the four of us. We had showers after more than a week of zero running water, put on makeup and went out for dinner on the town. People throughout much of the Mediterranean and Europe generally didn't even think about getting ready for dinner until 7 or 8 pm when the towns and villages came alive with chatter and storytelling over slowly drawn-out meals. The eating and occasional glass of wine wouldn't finish until nearly 11pm. It was always warm after all, and everyone was so inviting.

There were fountains in most if not all town squares, and families with children enjoying themselves at every turn. It was here amongst the small villages with goats and olive groves scattered around, with my British friends travelling up hillsides to various towns in Greece, that I felt the exact moment of release of the pain from watching my first boyfriend try to kill himself. I had struggled with the guilt and the heavy weight of not being able to control anything about that experience, and the pain he must have been in to cause him to do such a thing. It was sharing time and pleasure with some very warm and inviting people in a breathtaking atmosphere that helped me to let go of the sadness, anger and pressure I had placed upon myself. We spent a few more days together before their time was up and routine life called them back to the United Kingdom. I was sad that all my new British friends were leaving. We exchanged details and they invited me to stay with them in the future if I felt it was in my plans. After they departed I cruised around more islands by myself always heading further south, and spent some time on Crete. I thought it would be warmer on that island, seeing as it was more to the south, but it only turned out to be much windier. It was definitely no longer summer in the southern Mediterranean.

While walking around the Fortezza Castle in the town of Rethymno, I made fast friends with two Italian men who just thought the world of me. They were a little older than I had been at the time, and worried for my safety. One of the men took a bit of a shining to me as we walked and we took a photo together. He said he was going to tell everyone back home I was his girlfriend and they would all be so elated that he was able to find such a gorgeous Canadian woman in Greece. I thought it was hilarious, and did not feel in the slightest as though my safety was in jeopardy due to his comedic advances. They may have been military men, although I can't be sure, but I do know they were very athletic. We spent a few afternoons together wandering through town, and despite his consistent proposals to be the love his life for all eternity, I declined and they continued on their journey.

I had been staying with two women from France that I had met leaving a transit station in a small apartment in the city, and they were able to direct me to a laundromat that cleaned my clothes by the load for very little. It was nice to finally have fresh smelling things to put on my body. The women were hospitable yet quiet and shy, and mostly kept to themselves. I felt awkward being around them most of the time as we did not share the same willingness for idle chatter or for outgoing adventure. They wore one piece bathing suits and did not shave anywhere, as I found to be quite common with many people I had met from the same region. I felt as though I was becoming a more diverse person simply because of those I met, including these two women. We were together for two to three days before I decided to continue on my own again.

I drifted to the southern coast of Crete and followed my guidebook to a small and cozy hostel in Sougia. I preferred staying in Hostels while I was travelling alone as it was the easiest way to meet people, make connections and share experiences when there was no one else to share them with. Generally speaking many of the people I met in hostels or penciones were of the like mind that sharing, experiencing and connecting with anyone added so much more value to the experience of a place or a monument of any description. We were all there to help one another and many, including myself, would often leave behind things that might have been of better use for someone else whether they be books, maps, floss, clothing, or hats. Virtually every hostel I stayed or lived in felt like a community within itself. I met an older Canadian couple in Sougia and they spoke of a hike they had just returned from that would blow my mind. Unfortunately I did not get to see the Samaria Gorge that time around but it is on my long list of things to complete that I did not end up doing during my meandering journey through Europe.

The locals on Crete like most other places often expressed their worry for me when they learned I was travelling on my own as a young blonde lady. The men did not understand why I would carry such a heavy backpack, and looking back, I'm not totally sure I knew either. Many things I carried I didn't even use, or would have had a use for. Part of me wonders if the weight I made myself carry represented the suffering I felt I needed to endure because of the hardships I had faced, or the pain I thought I had inflicted on another human being. I felt the weight of wondering whether or not I had been the reason someone had tried to end their life, even though I knew deep down it had nothing to do with me. At the very last villa I stayed while on Crete I regularly saw the manager of the complex looking from afar at the windows of my villa to make sure I was okay. It made me feel uncomfortable, however as I didn't want other people to think I needed rescuing or assistance. This, coupled with the incessant wind and consistent cooling of temperatures, persuaded me to return to the mainland.

I caught an overnight ferry back to the bustling port town of Pireaus, cutting my accommodation budget as thinly as I could by doing the travel and the sleep together, although there was never much sleep to be had when I tried to do both the travel and accommodation together. I don't know that I would choose that option again in my now brittle old age where I turn into a pumpkin if not for my ultra-plush mattress. As we were getting deeper into fall, there were more and more migrants on the ferries so there was less space, quiet, and privacy. Everyone on board bought the cheapest walk-on tickets and congregated in all of the shared spaces to sleep.

There were large extended family groups of 30 or more in the theatres, under the tables in the food halls and along the corridors carrying small children and bags made from woven plastic. They had literally brought their entire lives, or as much as they could carry with them on the voyage. I was well aware what their intentions were by being on this boat, in the middle of the night, in the middle of the sea and I kept trying to figure out whether the majority of a certain group was Romanian, Albanian or other. One of my favorite hobbies then, and that I still practice to this day, is to listen to other languages to see if I can correctly pick out a dialect or a home country. I am not always right, but I do enjoy trying.

Athens was somewhat less overwhelming to me this time around and I went to a Hosteling International recommended hostel in a neighborhood in a northern suburb of Athens. I spent a couple nights there before going further into Omonia, after following a couple I had met at a market to their hostel. I had heard through the backpacker grapevine that this place was much cheaper with a nightly rate of about 11 drachma. It was here at the hostel that I met a very friendly group of castaways and decided to apply for a working permit to travel and work in the United Kingdom to meet up with the collection of friends I had met earlier in the Cyclades.

I settled in quickly and began introducing myself to everyone at my new temporary home. I even organized a toga dance party for Halloween set to a local music video station on the television. The manager of the hostel eventually offered me a job working at the front desk and provided me with free rent and some spending money after he had heard I was basically stranded waiting for a visa. I worked alongside a young Japanese fellow who showed me the most technically advanced laptop software I have ever seen even until now, some 15 years later. He was quiet and respectful and taught me how to manoeuvre through the rotating file storage icons he had on his laptop. I remember thinking the graphics looked like a moving DNA strand with sub-files within files, and it was a much more visual concept than anything I had seen. I made very fast friends with a Malaysian man who was older than me but who took care of me as though I was his sister. He also worked behind the front desk when I did not. He was soft spoken and had straight hair cut just above his ears, and very sharp features.

Working the front desk of this Hostel was quite the experience for me, wide-eyed and largely innocent at 20. Across the street was home to a brothel where humans were trafficked and the police were always around. I heard gunshots on more than one occasion, and we were trained to beware of prostitutes trying to come to "stay" at the hostel for varying lengths of time. We were all very protective of our home, however makeshift it was, and dealt promptly with those who had other intentions.

It was a simple older building with many rooms with numerous single beds and shared bathrooms on the ground floor all furnished simply yet comfortably. There were a few private rooms as well, although I never saw any of them. I noted that most of the showers throughout Europe had showerheads centrally located in fully tiled bathrooms and lacked shower curtains, which made shaving legs by placing my foot on the lid of the toilet very easy. There was a little one room cabana out the back in a small courtyard where my Malaysian co-worker stayed. I really felt homey and welcoming, despite it's rough edges and lack of modern conveniences.

The hostel had set up a few small camping stoves and we bought fresh chestnuts from the open air market near Omonia Square and roasted them at night like popcorn for all of the guests for a little extra warmth. They were so delicious and fresh. The Olympics were about a year off at the time I was there and the city was scrambling to complete the Metro in time. There were construction crews all over the city trying to hurry and fix a lot of the historical aging buildings and infrastructure. Strike action was so commonplace I barely batted an eyelash, as every day I saw another group of people marching along a roadway, blocking traffic and proudly waving placards in the air. The country had yet to convert to Euros, but the impending switch was all anyone was talking about. Most travellers worried about how that would change the cost of their holiday while locals worried about how it would impact their livelihoods as tourism accounted for the majority of incomes on the islands. There was an edge of uncertainty yet life somehow continued to pull forward regardless.

I went weekly to sit at the British Embassy to check progress on the paperwork. It seemed so much simpler and rudimentary back then. Nothing was electronic and my passport was actually mailed places in order to get the appropriate stamps and stickers. I went to downtown Athens and visited museums to pass the time. Duran Duran even played a concert at a major venue, for which some t-shirt scalpers that were staying at the hostel gave me a free concert shirt. This I learned was something that people did, travelling from country to country selling bootlegged concert tour t-shirts at a fraction of the price outside of the venue so instead of paying $30 for the official one, they sold them for $5 or $10. They came with hundreds and left with very few.

One day I walked to the National Gardens in downtown Athens and sat under a tree to read a book. About three pages in, a man came by and sat on a park bench about 20 feet in front of me. I looked up and made eye contact with him, and settled myself back in to reading not thinking too much about seeing another human in a public park. A few moments later I noticed movement coming from his direction. I looked up to see that he had taken out his private parts and had begun to pleasure himself in my presence, and that was the end of that. I quickly scooped up my book, gave him a disapproving look and headed back to somewhere with more people and less privacy. I thought I had seen it all at that point, only to be proven wrong. I told my park incident story to a few people back at the hostel and all of them told me half joking they would not let me go out on my own again.

Once my passport was back and my new visa had been added to its pages, the Malaysian man and I decided to travel together back to Italy and I felt relieved to have a travel buddy. He didn't speak much English but we got along nonetheless. Our Mediterranean sailing had gone smoothly, however he was detained by the harbor border patrol when we disembarked the ship in Italy. He'd said immigration was always sticky in Europe with a Malaysian passport and he told me to go on without him and we would meet up once he had made it in to town. I never saw him again and have wondered about him often.

I travelled back from Brindisi to Rome by train. The leaves were a rainbow of fall colours from slightly yellow to vibrant red and coloured the passing landscape like a classic oil painting. The only issue I ever had while travelling happened on a train in Italy. The conductor must have thought I was lost and noticed I was travelling without companionship. He came to help me carry my backpack on to the right train car, going to the right destination. He was a short, stout middle aged man who did not reach my height. He was covered in sweat and thick as a possum greying facial hair. He picked up my purple backpack and showed me to a private room on the train. There were not many people on the train at this time, and most had already filed in and nestled themselves in their own bunk rooms. He hoisted my bag above the seats in the overhead racks and turned to me. In Italian he started what I can only think in his head was some soliloquy to woo me. I could make out bits and pieces "your hair is the like the wheat in the fields, your eyes-shine bright like the stars in the night sky, your skin...." and his hand came out and caressed my cheek. In this moment I felt as though I was very rapidly descending into a bottomless pit of fear. He grabbed my face with one hand and my body with his other and pulled me very tightly against his sweaty body. His prickly moustache and beard scraped up against my face unapologetically. And a moment later I felt his slippery warm, moist tongue plunging down my throat. I pushed him away with both arms, stood there with a shocked and somewhat confused look on my face, grabbed my bag and fled down the now-moving train to find two even more blonde Swedish girls a few train cars forward. I let myself into their cabin, threw my bag up on the racks, locked the door and declared they had a new friend for the rest of the trip. Once they heard what had happened they were sympathetic, and rather aghast. I wasn't all that disturbed and shook it off quickly once I knew I was in good company. Swedish women knew how to throw down. Other than making out with someone I would not normally have, not much bad came out of it-as long as we all made it to our destination unscathed. And we did.

I ended up back to the Yellow Hostel in Rome from where I first began my Mediterranean adventure. They were surprised to see me and wondered if they were going to have to throw out the luggage I had left in storage more than two months earlier. From the time I first stayed at this place until the point of my return, the two cute young hostel workers went from not knowing each other to being heavily coupled. It was a very cute thing to see two strangers in the length of time I was out being lost, come together and find solace in each other. I rearranged my bags and laughed at some of the items I had packed for my initial travel plans. Hair dryers and high heeled shoes that had absolutely no purpose now that I was officially a European hobo as opposed to a high class nanny. I was very excited to be on my way to England, to meet my island friends once more.

CHAPTER SIX
BRISTOL, THE FIRST

I really had to do some wrangling when I was back in Rome with the travel agency. My previous flight ticket did not permit me to have an open jaw stopover in London, which is where I wanted to go for an extended period of time. I didn't want to spend my entire flight back to Canada but was finding it difficult to break up the legs on the very concrete paper ticket I held. It took about two days of repeatedly going back to a specific travel agent and speaking to the only employee that spoke half-decent English numerous times throughout the days to see if she had figured out how to skate around the logistics of my ticket. At last she contacted the Canadian sister equivalent of her company and had figured out how to get me to London. I had a fresh Visa in my passport, and was full of excitement to be surrounded by buttery British accents, and earning pounds. My financial resources had gotten me this far by working odd jobs in odd countries. I had nothing, yet it was enough. I had largely gotten by from peoples' kindness and generosity and the trust that everything was going to work out just amazingly. I never once questioned this haphazard, ill thought out plan. I just knew it would work.

I love flying European airlines as they all have their own distinct feel, and I always loved comparing uniforms, meals, snacks, hospitality, and leg room. When I arrived at London's Heathrow for what would now be the second time in my life, I was fairly overwhelmed. Heathrow is basically its own town, with stoplights and shuttles, tube stations and countless terminals. Getting shuttled from one place to the next was a maze, and finding the right bus to get me to my new destination of Bristol was rather complicated. I eventually made it to the bus station with row upon row of parked buses—or Coach lines as they are commonly referred to across the

Atlantic. Coach lines in England were a treat to be sure. During my journey to Bristol a young man would walk up and down the aisles selling chocolate bars and crisps, pop and cans of juice or alcoholic beverages every 20 minutes or so. It was now November and much cooler than the Mediterranean. I had three layers of sweaters on and my only pair of jeans, which happened to be my favorite. They were dark denim Miss Sixty Jeans, the kind that had the zipper up the back of the ankle, slung low on the hips, and just enough stretch to feel like a second skin. I did not have a jacket, and felt the frosty frozen air on my skin all too much.

As the coach cut through London, I caught a glimpse at the lights and sights of a very large, very busy metropolis. Everything smelled like diesel. Something I noticed about most European cities was their lack of tall sky scrapers I was so accustomed to in North America. The sprawl was immense, but it was nice to be able to see sky, and to not be walking in shadow all the time. As we came to the city of Reading just west of London, I wondered hopelessly and haplessly if by some miracle of miracles that young man I had spent a day with in Rome would jump on the coach and we would live happily ever after. I kept his name and wrote it on the back of a photo of him in the lift at the Yellow Hostel. Looking at the photo now, I wonder if he is still alive and how he has spent the last fifteen years. Alas, he did not hop on that precise bus, at that precise time and I have not seen him since that day.

When I arrived in Bristol it was late evening and I was STARVING. Elizabeth and her boyfriend picked me up at the coach station in town and welcomed me with open and warm arms. We went to a tiny Indian restaurant that cost 5 quid (Pounds) for a traditional steel divided plate of food. It seemed like the front room of someone's house, and there was only one type of meal on the menu. We had to stop at a bank on the way there because I had no local currency, only my debit card. I never got travellers cheques or exchanged money before going to a country. I almost always would just withdraw from a local ATM wherever I went, and I still do this today. I hear the exchange rate is terrible by doing this but it is a lot more convenient than finding a foreign exchange kiosk.

We sat in a front room somewhere in Bristol on a very cold early winter night and I filled my belly with the most delicious dahl lentil, chick pea curry, naan bread, steamed rice and anything else I could get my hands on. I cleaned my plate and was still hungry afterwards. We drove back to their flat and chatted lightly on the couch for a while until I slid soundly asleep, not knowing what or how I was going to find work, or find my way around. I was hungry to be employed and I felt a deep need to be able to support myself, pay rent, buy food and get my hands on some warmer clothes!

After what seemed like only a few days I had made a whole band of new friends through those that I met on the islands in Greece. They only lived a few blocks away from each other and I found it easy and comfortable to make friends. I went for countless interviews in my one pair of jeans and my one sweater. Finally a very young Scottish manager of a bookstore in the heart of the downtown shopping district offered me a job working with him and two other lads about my age. A young budding musician by the name of Andy and I became friends very quickly. He liked electronic music and would make me CD mixes of cool British and European groups which I regularly popped into my Discman to listen to on the way walking to and from work every day. I also received a hot tip for a job working as a bartender on Corn Street at a place called Edwards through one of my new Australian friends. This would be my first real bartending job, and also a crash course in British pub culture.

Before I had started at the bookstore, I did my first few shifts at the pub Edwards. Carlos was my manager's name. He had typical British teeth and smoked a few packs a day of some British brand of cigarettes. The ashtray in his two square foot office was littered with ashtrays that were stacked to the brim with crushed butts. He gave me a uniform top, which was a shiny black V neck top with a small emblem on the left shoulder and a key card for the cash registers and told me I needed to buy black dress pants before 4 pm that night. I went up the road and entered the first shop that I thought might have such a thing. I grabbed the first pair of black pants I saw, in the wrong size because I had no idea at that time what a UK size 12 was and found the change rooms. Change rooms in this particular store amounted to one large open room, with one large floor to ceiling mirror covering one side. There were about six or eight women with their mothers and friends, all stripping off to their "pance" (underwear) and other underthings and openly critiquing themselves and the clothes they were trying on. I was pretty shocked at this, as it was in such stark contrast to the modesty of Canada. I had never even felt comfortable changing for gym class to full skin in front of others. I did have a few months of nude sunbaking in the Mediterranean under my belt at this point though, so I actually felt quite comfortable after about four minutes of trying not to look at all the others.

It was another thing to hear them all talking openly about one body part or another, not in a harsh or negative kind of a way, just as matter of fact statements. Everyone giggled now and again and it was a light and cheery atmosphere. The black pants I tried on were obviously enormous on my cricket, boy like skeleton and after figuring out sizes a little better I settled on a wide legged, double belt-looped pair that called for huge hips that I simply did not have, but I needed to be at work for my first shift in England in less than 2 hours. I bought the hip-full pants and went back to my new home, changed, put make up on and ran back downtown. It took about 40 minutes walking swiftly by foot each way and for some reason in the entire time I lived there I never once took the bus.

My first night in my first bartending job went smashingly, at least I thought so anyways. I was put right at the front in the middle of the U shaped bar, between two very seasoned employees, and was basically told to have at it. I hadn't had the foggiest idea what the British coins looked like, I had never heard of any of the brands of beer before, I hadn't even ever poured in to a pint glass before, but away I went. I started taking orders in my corny Canadian accent, I searched for whatever it was they were looking for, learned how many ounces went in what, learned where the glass washer was, and did pretty okay! I didn't break too many things, and people were pretty quick to correct me if I had given them the wrong change. I learned quickly what "After Shock" was, and how popular "Reef" premixed drinks were.

The steepest learning curve was for things like "Carling Shandy's" and "Blackthorn with Black." "Tew pints of Blackthorn me luv" I'd hear. My new black pants were drenched with beer and sticky liqueurs from pouring shots. The floor was ankle deep in bottle caps, and the glass bottle bins were full to the brim with empties. The seasoned professionals patted me on the back, and said I didn't do so badly for someone who had never seen much of the money or had ever truly slung a pint before. I was elated! Great success! I'd even made a few tips for my troubles by those that clearly thought it was "cute" that I had no idea what I was doing. After a month or two, I even made it in to the local paper, for a story that somehow linked going out for a dance and having a drink at the pub as actually healthy for you as it seemed like exercise. It failed to mention the side effects from such a dehydrating workout, but I got a kick out of having my photo in the news.

During the days I worked at the bookstore from 8 to 4 and the pub from 5 to 2, I'd make the 40 minute trek home in the pitch black, usually with sheet rain flying sideways at me, coating my face and soaking my clothes. I'd wash the rain off my face, brush my teeth and fall in to bed only to be woken up hours later to start the whole cycle all over again. Unfortunately due to this hectic work schedule I very much was not taking care of my body in any shape or form. My newly acquired belly button ring that had been placed in my abdomen a few months earlier amidst the lush green hills of Greece was now going green itself. I found the ball on the top of the metal stud would often catch on the plastic rim of the glass washing racks that I loaded and unloaded every night in the kitchen of the bar I worked at. It had caught so many times, and began to get infected. I found my stomach was always hurting because of this and eventually had to remove it in order to let it heal properly. Luckily for me the infection left my body the minute I slid out the piercing, and healed over not 3 days later. I don't think I could ever have cosmetic surgery done, my body just doesn't like being tampered with. I didn't mind the crazy and hectic working schedule at all. I knew I had been on a few month long holiday beforehand and was very willing to work hard to continue having more opportunities to get back at travelling. My credit card which had a limit of $1500 CAD at the time was sitting at a balance of about $1200 from travelling, which was the most debt I'd ever had at that point in my life. On the days I didn't work at the bookstore, or on Sunday and Monday evenings off from the bar, I would cook stores of food for the week, do laundry and chat with my friends. It was a very simple and humble existence, and it suited me just fine.

After a few months I acquired a winter coat from the back corridor of the mall the bookshop was in, as one of my coworkers James had found a whole bunch of thrown out odds and ends from a retailer. My manager gave me Jamie Oliver cookbooks and a skateboard for Christmas, and Edwards threw a fabulous gathering where everyone cut loose and danced for hours. Inside the bookshop over the holidays like any good retailer, we played one CD of music over and over again. The playlist was burned in to my brain and I still remember every song, probably in order to this day. Every time I do hear the songs again I am transported back to that time to that place where we would stop every afternoon for a McVities biscuit with chocolate on one side and a cup of tea with milk and sugar. It was absolutely splendid and oh so British. I have tried to search for the name of the chain of this Scottish bookshop but have never had any luck finding it. I often think of the very small group of men I worked with. I feel like we were all so young then, and so clueless.

It was at about this time that the relationship between my girlfriend Elizabeth and her partner had really begun to sour. One night after I had returned home from work at 2 am, exhausted from another 18 hour day of straight work, her boyfriend stumbled in drunk from a night out with his friends. He crawled in to my bed and began aggressively man-handling my body. I was completely unconscious and remained that way due to sheer exhaustion for a few moments before I awoke to this person who was much stronger than I, having his way with my body including my most intimate bits. I was so tired I just kept asking him to get out of my bed and go find his girlfriend. I let my girlfriend know the next morning and she was less than surprised, she let me know he had done such things before and regularly got so drunk he "didn't know what he is doing."

He moved out shortly after that and I noticed on my chequebook someone had written cleanly and neatly in the top corner "gaygirl". I had no idea what it was about at the time but it came out years later that he had thought that Elizabeth and I were actually lovers and that is why she broke up with him. Not because he treated her terribly or tried to sleep with her flatmate, or used drunkenness to excuse behavior, or any possible other reason. One evening I went out with a now newly single Elizabeth to a pub for some drinks, laughter and dancing. We had dressed in our finest and put on black heels. She spent an hour doing a fantastic job on my makeup and we proceeded to have a very fun night out meeting her friends and manager at the time. She flirted up a storm with a couple of handsome men, and I flirted as well. On our walk home in the crisp winter evening I began to lose the ability to stand. We were laughing and carrying about with two men who had followed us out of the venue and insisted on walking us home. I started not being able to make sense and thought I had simply had too much to drink. My legs gave out from under me and I stumbled to a curb while holding on to Elizabeth's arm. I can't say I remember clearly what happened for a brief moment but I do remember we ran in to two of our friends who also happened to be walking home and intercepted us, completely wasted on the side of the cobblestone road.

The two guys that had followed us quickly took off in another direction and one of our male friends had to pick me up, throw me over his shoulder and proceeded to carry me all the way home. I was violently ill for the rest of the night and most of the next day. My girlfriend disrobed me, showered me, redressed me and tended to me for days. I had no idea how I had been so sick and she had surmised I had been drugged. I am so incredibly thankful our other male friends had intercepted exactly at the moment my legs gave out. I was careful when I went out after that. I learned to not leave my drink unattended and always watched other women's drinks while I tended bar to try and cut down their risk of having the same thing done to them.

It quickly became springtime in England after a few months of getting my feet grounded and enjoying being able to have my clothes cleaned in an actual washing machine in the kitchen, as opposed to washing everything by hand in the sink as I had done for months prior while meandering through Europe. I began to feel it was time to go off again and explore. I went on my first official British surf trip to a small coastal town in Cornwall called New Polzeath. It was an idyllic cove along the southwest coast of England. I went with a small group of four friends and camped in an old restored VW combi van that we parked in a field near the coast. We ate fresh seafood and swilled crisp lager at a pub that was perched up on a hill overlooking the shoreline.

I borrowed a wetsuit and a longboard from my friends and headed out to the surf. I remember the first wave I caught and rode all the way to the shore. It felt like it lasted five years but was probably 15 seconds. I was so full of excitement in that moment as I slid along the top of the water after hours of pounding against the break with this massive surfboard. It was so hard for me to control in the water and wind, but the length made it easy to stand and ride to shore once I had made it out that far. We spent evenings getting warm in the local restaurants and played cards in the van in the early mornings while waiting for the mist of the morning to burn off.

I also went often to Devon with Elizabeth to visit her mother and siblings. We hitchhiked across the county at her insistence. It was the first and only time I had ever done such a thing and was actually extremely fun. Within about 10 minutes of the two of us standing at the side of a highway near an on ramp we had nearly a dozen or more cars stop all at once, and almost caused a multiple vehicle accident when they saw our hand drawn "Plymouth" sign. We ended up riding with two men from London who had a drum and bass studio where they recorded music for guys that I'd only slightly heard of through either my flat mate or my workmate. They drove a BMW, which I am pretty sure is why my girlfriend who was in charge of the hitchhiking mission chose them. However rough around the edges they were, they treated us kindly and with the most respect. My friend somehow convinced them to go completely off their course in order to deliver us to the village we were trying to get to, and they obliged without hesitation. They spoke with my friend for much of the way down about music and melodies, collaborations and mixing equipment as Elizabeth was a singer and songwriter and they had a lot in common to chat about.

Elizabeth's mother had a darling cottage at the edge of a small village where we would congregate and chat for hours over a pot of tea with clay mugs made by hand. We would laze in the afternoon sun and give trims to each other's hair while someone would strum melodies on a guitar as we snacked and napped. On one occasion we stopped at another friend's place in the country. He owned falcons and owls and all kinds of other exotic things. He had a big netted trapeze hung high up in the trees that we spent endless hours playing in. There was also a pond on the property where we would row a small boat and try to tip each other into the freezing cold late spring water. It was a beautiful time, completely carefree. Sadly my girlfriend's sister has since passed, in a rather tragic and unfortunate circumstance. And I miss her dearly, as all our interactions were always so full of love and lightness. I still cry to this day thinking about her. So many lives have gone away before their time in my opinion. I try to remind myself to be grateful for the time I did have with each one I have lost during the course of my life.

Throughout the spring I had loosely been planning to go overseas again, and had saved some money from my crazy working schedule to afford to do such a thing. One of my friends had given me an old mobile phone which I had been using as a pay-as-you-go that would work in every country. It was a blue Nokia 5150 that everyone had. It was a sturdy thing and I kept it with me through countless countries and continents. It withstood being dropped repeatedly and was much more rugged than anything available now. I didn't really have a clear plan of where I was going, or even where I wanted to go or end up, but I went anyways. My girlfriend let me keep the majority of my things at her house which I appreciated. I decided I would head off that spring with my first stop being across the channel to France.

CHAPTER SEVEN
PARIS

Now one would think being a young woman travelling on her own with some experience that wasn't necessarily the best, I would be reserved and cautious. I probably shouldn't just open up to total strangers out in the world and strike up conversations, letting them know I was just out by myself for who knows how long, going somewhere or other. You'd probably think it would have been best to keep mum about the fact that no one would probably miss me after too long, but apparently I didn't think that way. I totally trusted the life that was set out before me. I skipped lightly up the stairs to the coach bus that was taking me across the channel to France. I didn't even have lodging confirmed in Paris for my first night there. I had my Lonely Planet guide, and fully trusted it would all work out. There were scads of choices for hostels and auberges. I had absolutely no way of knowing for sure that everything would be fine, but I instinctively did despite the thousands of reasons it might not go easily.

I walked onto the bus with my purple backpack safely stowed in the underneath compartment and a small bag for my travels. I'd really only packed the essentials; one pair of jeans, two dresses, a pair of shorts, a bikini, a toothbrush, a comb, my passport, debit and credit cards. Basics. I made it a few rows deep and decided to sit beside this young French guy who happened to be reading and giggling over American Psycho. Hmm, great choice--methodical and sound. Of course I put my bum in the seat and flicked on my Discman to the sounds of Manu Chao and Nightmares on Wax.

About an hour into the journey, with the sun still high in the sky and streaming in the window, I took an opportunity at a pause in his reading for laughter to ask him about the book he was reading. I had only heard about it at that point, hadn't read about it and I don't even think the movie had been made yet. He spoke in fairly decent English and told me he was a student that had also been working in London for the good part of a year. He was looking to move back to Paris that following fall and land a big job in management for some large company in finance.

He practiced his interview questions with me, and explained the art of the interview and different techniques in how to make even your weaknesses your strengths, and your strengths part of weaknesses. He was on his way to stay with his brother for a few days in one of the arondissements in the city. He asked what I was up to and I loosely explained that I was off to see more of Europe. After chatting for a few hours about culture and living in the United Kingdom as a foreigner, something both of us had in common, I mentioned I hadn't really had anywhere all that solid to stay. He texted his brother who apparently did not have much of a problem having some random petite Canadian girl stay at his flat. It was miraculous and wonderful.

When we arrived in Paris he gave me a run-down of all the different areas in Paris, cultural things to see, restaurants, the Seine, museums, the most popular boites where all the young and ultra-hip cool cats hung out. He took me along the metro to his brother's neighborhood and we went to the local village shops where he picked up the finest, tastiest cheeses, freshest baguettes and a bottle of champagne. We went back to his brothers flat, which was open, airy and bright (also quite small and super expensive), and he laid out this most perfect platter of what he called the "true Paris". The curtains were white lace in the flat and the floor a very traditional parquet wood. There was no shower, only a bath with a spray nozzle on a hose which thankfully I was used to from time living in Montreal. His brother seemed somewhat confused at my presence but thankfully went along with it regardless.

We met his friends that night at an ultra-trendy boite for drinks and snacks. None of his friends were surprised he had collected a friend on the bus, which seemed off to me as he really didn't seem like the type at all. The boite was lit dimly with red light fixtures. The space was long and narrow and quite loud with chilled out jazz music streaming hazily through the air. His friends were lovely, warm, welcoming and generous. I could not have asked for a better first night in France. He took me walking the next day to museums, chapels, gardens, cafes. Tuileries, Luxembourg, Champs Elysees, the Louvre, Montmartre, and the Arc de Triomphe. We sat and had baguettes from a street cart with pigeons flocking around us and thousands of tourists snapping photos of everything and anything.

That night he even took me with him on what must have been the absolute most awkward date of some other woman's life! He had told her beforehand that I would be the third wheel in the dinner event, but it was still strange. I asked if he was sure, but he insisted. She was less encouraged but again totally went along with it. He wanted me to experience the cheese district so we went to arguably the smelliest neighborhood around. There must have been over two dozen restaurants in this neighborhood, all surrounding a central courtyard complete with a fountain covered in dozens of grey pigeons chattering with French accents (or so I'd imagined), all with menus full of every type of fondue you could possibly imagine.

When I stop to think back to that night I can still smell the pungent aroma wafting from every direction getting sucked up my nostrils. I was quietly cringing inside but he was more excited than a toddler on Christmas morning to fill his mouth with soft, liquid cheesy gooeyness. His date was a lovely wavy-haired, dark-skinned woman who was trying so hard to not continually look me up and down wondering what the hell I was doing on their date. I can't blame her. I just tried to make nice and have a non-threatening feel about me. I absolutely had no interest in this man in a romantic relational sense. We sat in one restaurant up a few stairs on the second floor that had white and red table linens, drank wine and ate appetizers, and then moved on to the next. It was just dusk in late spring so probably around 9 pm as we moved on to the restaurant where we would have our main dinner meal, and another few bottles of wine. Meals had such a more pleasurable feel to them, nothing was rushed and no one made a hurried lunge to get a cheque. By the end of the night they were canoodling and kissing and she had come to peace with my very odd presence.

I spent a few days riding the metro exploring different parts of the city by myself. I walked along the river and tried to see the romance in the scenery, but I don't think it is my brand of romance. Everything was cobblestone, concrete and pigeon poop. The wind was chilly and the trees hadn't quite fleshed out to full summer foliage. The river was greenish grey, and flowed fluidly along the concrete banks. I don't know, it could have been the lack of company in not sharing the experience with someone, much like a tree that falls in a wood with no one to hear it. I was mesmerized by the trains that were double-storied and fast as the dickens as they followed the Seine. Public transport and bicycle riding was much more a "thing" I felt here than in my experience growing up in North America.

My gracious host took me to a house party to see some more of his friends the next night in an even smaller flat in a university area of the city. The apartment was no more than 300 square feet and considered lavish to the two or three students that made it home. We nibbled on platters of charcuterie and cheeses that one of the roommates had brought home from his catering job. We drank wine from small table glasses and mugs. They were so warm and embracing. The typical two cheek-kissed and hugged on entry and exit. They told me stories of the Parisian mind set and how they decided to move to the city from various towns and villages throughout France. They spoke of the equality and bipartisan ways of those they knew, and how even though Paris was not affordable and extravagant, it was a warm and inviting place, brimming with chaleur.

On my last day I thanked my hosts as graciously as possible, as I had made a weak reservation at a nunnery in Bordeaux. I made off with my purple backpack across town to the bus depot. I took the very slow meandering path through the city and spent some time tracking down the Eiffel Tower on foot. From far away it looked like a massive marvel monument. You could see it from miles away, which made it easier for me to find in the days before mobile GPS and SIRI to help guide the path.

It took me the better part of three hours to finally get to the base of the Eiffel Tower. I had no map, no idea what the city layout was, and a 40 lb. bag strapped to my back. The Eiffel Tower is actually surprisingly way smaller than you would think it to be up close and personal, or at least I thought so. I was so unimpressed I didn't even bother to take the ride to the observation deck for some obscene amount of money. Instead I made myself a picnic at the base in the grass and joined a whole bunch of other lovely couples who were having picnic lunches they had carried full of baguette and cheese, fresh tomato and red wine laid out on red checkered linens. The sun was out, and it was a splendidly warm day. Not much of a breeze and not much noise other than the occasional ambulance in the distance. Beautiful.

Despite the literature my bus mate was reading and how it should have been telling me otherwise, I feel most fortunate to have had such a humble, gracious host take me on the most personalized tour of Paris. I had taken the fact I had no plan in stride, and it absolutely worked in my favor. I could not have asked for a more intimate experience, nor do I think I would have met the same people if not for this most unexpected opportunity. I'm not sure how I would feel to do the same now, knowing what I know and experiencing what I have, but I do feel like this was a most gentle introduction to my time in Europe. I felt deeply inspired. I finally reached the bus depot in the late afternoon and slowly began the drive out of the city to the south, always south. This time to Bordeaux or so I thought.

CHAPTER EIGHT
SPAIN

When I stepped on to the bus, which happened to be an overnight coach bus to Bordeaux, I thought I'd try a different approach to finding seat mates. In fact, seat mates beckoned to me this time. A group of three young travellers' mid-way back in the bus called out to me and asked me to sit with them. They must have noticed me sitting on my purple backpack at the station and thought it was a nice thing to make me feel less lonesome. They were a few years older than me, and incredibly nice. We spent the first few hours talking about where they had been and where they were going. I had said I was headed to Bordeaux, although I wasn't all that sure, and still am not to this day certain as to why I thought that would have to be my next stop. My thinking must have been wineries, and nuns, which seemed safe. I still hadn't really carved out much of a plan for the entire few months of travelling I had ahead of me. I realized that things would probably just continue to unfold as they would. In conversation with this small pack of travellers, they started speaking about where they were headed, which was Madrid. I knew nothing of Madrid, so I grabbed my Lonely Planet and started reading up on the city. A few pages in and I began to question whether or not Bordeaux was where I should have been going. These new travel mates were more than happy to have me tag along, and I had an open-ended bus pass throughout Europe so I decided I'd just carry on to Madrid.

When the Bus pulled in to the Bordeaux station somewhere around midnight, I suddenly felt very timid and shy. I didn't want to make a scene by having to go up to the front of the bus to explain myself, and I thought that if I pretended to be sleeping, then no one would notice and we would just carry on. We sat at that station for probably half an hour past when we should have moved along, and I was sweating profusely from the stress of not knowing how to handle anything, so I finally made my way to the front of the coach to speak with the coordinator of the bus driver (they were two separate people) and asked him in the kindest of ways if I could just continue on to Madrid given my essentially open-jaw ticket. The tall, dark haired stately man with a slightly soft belly hummed and hawed for a while, mostly for dramatic effect and then signalled to the driver we were to continue on.

I apologized for being so timid and embarrassed and he quickly ushered me with a wink and a wave of his hand back to my seat. I was pretty excited to be going to this new place, more than my Bordeaux ambivalence. We arrived in Madrid early in the morning and made our way to a café, and eventually to a very cute hostel near the middle of the city called Los Amigos. Madrid was humid but not hot. It was overcast the day I arrived, but the clouds burned off on our walk to the hostel. It was nice to have met these random people on the bus, and it made for the walk to find the accommodations much more chatty and pleasant.

I made friends quickly with others at the hostel and settled in to my bunk bed for a bit of a nap after the overnight bus ride. I was never one for upright sleeping. Neither m neck nor back never really did find a groove with that whole situation, even with neck pillows, or laying out on seatmates rather unapologetically. When I awoke sometime just after noon, I found my cheap (less than 10 euros a night) dorm room had one new guest, a beautiful brunette from Bordeaux, who wore thick colourful scarves in her hair like headbands. We chatted up a storm in the café area of the hostel over baguettes, butter and jam and made our way out to see some sights. First stop was Gran Via, a rather well known intersection of the main auto routes in the city. It was extremely busy with cars, buses, scooters, bicycles and pedestrians. The architecture was all brick and cobblestone, domed roofs and massive billboards. It was lovely to hear the Spanish chatter of the locals as we strolled to cafes and clothing shops. She was an absolute delight to have accompanying me, and it gave me a great opportunity to flex my French language skills once more. We visited the Prado, Plaza Mayor and Palacio Real. I found it interesting how much richer experiences were when I shared them with other people, even those I hadn't known for too long. Sharing a moment in time with someone added a dimension I didn't otherwise have when I was on my own.

A few nights into my stay at the hostel I met a young man named Steve who was from the Pacific Northwest in the United States. He was very cute, with light hair and blue eyes. He wore baseball cap and a t-shirt with printing on the front. We started chatting in the kitchen area while preparing dinner. I was making spaghetti with tomato sauce, and he was making sandwiches. We started by doing the usual backpacker bullet notes of who we were, why we were in Europe, and what we wanted to see and accomplish. The conversation slowly drifted on to other topics as the night wore on such as baseball and who or what we had left behind.

By about midnight we were sitting on the countertops once the rest of the travellers had finished their meal prep, and once a few groups had gone and returned after a night revelling at the sights of Madrid at night. Somehow we had drifted on to the topic of politics and the stigma surrounding American travellers in Europe. At the time I was there, most Americans were unapologetically sewing Canadian badges on to their backpacks as a way of garnering a little less hostility than the stars and stripes had allegedly been getting them. I cannot say for a fact whether this strategy was useful or even needed, but it seemed to be the trend at the time and we both wanted to talk about it. I remember this conversation so clearly because it was the first time in a long time I had formed an opinion about something that I felt I understood so much about; not about Americans per se, but perceptions in general. We argued rather amicably about American patriotism and what that meant and we spoke about why some felt it was better to be seen as being Canadian in Europe. I had absolutely no ill intent about the conversation as I was genuinely interested in hearing what he had to say about why Americans were viewed in one way or another to others abroad. I found it intriguing to hear the view of someone who had been faced

with a stigma, speaking about that stigma and how he felt ı
directly affecting him.

   At about four in the morning, still chattering away, him
vehemently upholding that it was everyone else's stubborn behavior
that regarded him and all his country folk as being a certain way
because he was from America. I turned to him, calmly looked in his
beautiful twinkling blue eyeballs and explained that this was exactly
my point. It was his thoughts that others saw him a certain way,
regardless of how he actually was, and his unwavering intransigence
on the issue that validated my point exactly. He couldn't even
fathom that there might be another way of perceiving anything other
than his own, or that we might be perpetuating the beliefs of others
without even realizing it. As in, because others think I am loud and
obnoxious, then maybe I am loud and obnoxious and so therefore
need to be defensive because I think people think I am this way.
When we connected on this point, on this realization together his
whole demeanor changed. And he stopped cold, sat back on the
counter and his eyes trailed the ceiling of the dimly lit room. Silence.
He put his hands on my shoulders, looked me in the eye with a red
face of embarrassment and gave me a hug. I honestly found myself
totally surprised that his face went red and he felt embarrassed in
that moment, I thought we had experienced such a breakthrough. I
just saw it as this pivotal moment in both our lives. He muttered "oh
my god" under his breath, hugged me again and left for bed. I felt
like that moment in time served as this powerful force of connection
with this other human being based on a thought, something I hadn't
felt in a long time, if ever to that depth. I walked off to sleep, not
even tired from an Olympic ten hour conversation, and awoke the
next morning at 8 am to find this man had left, without even saying
goodbye. I was totally perplexed by this, and saddened. I was hoping
to have more conversations of the same depth with him that day but
it was not to be. He was on his own journey after all.

   I carried on with my time in Madrid, somewhat lost after the
exchange with the man from the west coast of the United States. I

forget but it was something average and every day, like Jorge. I tried to squeal as quietly as possible in delight as he handed me Jorge to make acquaintance with. I held and petted this small scaly creature for some time. Green and tan with big beady eyes, looking up at me with indignation for having interrupted his nap on a warm lap in his little home. This young man was travelling home to visit with his mother and father and had been living in Madrid for some time. The Catalan accent almost sounds like a lisp on certain words, and definitely added more flare and drama to an already expressive language. He let me know some things I absolutely had to see while in Barcelona and gave me a quick rundown of the underground metro. He blessed my journey as we headed off the bus. He was a very kind soul, so much so that my short visit with him left a long lasting impression on my experience.

Barcelona is an incredible city. The streets are lined with lush green trees dotted with cafes and laissez-faire eateries. I visited museums, monuments and malls, soaking in the language and genial hospitality of the city dwellers. I spent hours at the various Gaudi sights and several days at the Sagrada Familia, scrutinizing every single speck of dribble of concrete on the original side that was done by the man himself. I imagined the thousands of hours it would have taken him to create each square foot of this grand monument that is still yet to be completed. The clear vision he must have had, right from the onset to what this would look like, is unimaginable. The level of detail for which this monument was created boggled my mind. Everyone in Spain naps in the afternoons and everyone suppers late in to the night when it is cooler and the winds are gentle, light and refreshing. Laughter echoed from every direction in the city. No one looked stressed.

My Montreal family, whom I feel already saved me once, had coincidentally decided to holiday with their family while I was there. We met in Barcelona, where I had found a downtown hostel to stay in. It was a fine place, nothing amazing, quite loud and boisterous and right downtown off Las Ramblas, and a super party location. Once they arrived I feel as though I was in paradise. Their cousin who lived in the city offered to have me stay with her, at no charge, in their beautiful two bedroom flat with free hot water and luxurious shampoo. I had previously been using 2 in 1 to save on luggage real estate and it made my very thick, wavy hair sticky and tangled. She was in a classic and typical European apartment complex with an internal courtyard where you could see everyone else's back balconies and washing hung out to dry. You could catch all the gossip of the building, and if you listened closely you could even catch the 5 o'clock news from across the courtyard although no one would do so out of respect.

My Montreal family took me with them as though I was one of their family to all the best restaurants, cafes, sights, sounds. I was so unbelievably grateful for the experience of five star restaurants in castles with endless bottles of wine and proper silverware. We travelled for a few days down the coast, eating in elaborate restaurants in caves with candelabras and seaside views. I met more of the family, who greeted me with open arms, fed and watered me until I was satiated and relaxed. While we were sitting in one of these opulent cave restaurants, eating the eyes of a fish, with a violinist playing accurately and lazily in the background, I started getting texts from a friend of mine from back in the UK who had decided to surprise me and come down to meet with me to travel together. We had probably easily crushed three or four bottles of wine at this point and my Montreal girlfriend had started replying to his text messages, warning him that there would be blood spilt if he treated me with any kind of malice.

It felt really nice to have someone do that for me. I felt loved, and pretty intoxicated. I felt terrible at the way I had to leave Montreal as I felt it was a choice made for me, and not one I had been ready to make. But alas, I was so overjoyed at having met up with them again half a world away as a bit of closure and reconnection, and I enjoyed their company in a very relaxed setting.

I left my Montreal group to go off to search for my UK friend who was en route from England. We had made plans to meet back in Madrid in the train station but quickly decided it would be easier to meet down in Seville, further south. The train I chose to take was slightly statelier than the regular milk-run trains I had been taking previously. The ticket was 15 or 20 Euros more than a regular fare, but the train was much newer, nicer, cleaner and faster. The entire passenger manifest consisted of young college kids and high schoolers on their way home for summer holidays, and others going specifically to take in the Flamenco Festival that would be happening in Seville. It was evening on the train, and the Spanish young women were incredibly beautiful. Some had brought their costumes for the festival; lavish flowing multicolored skirts done in dozens of tiers of cotton and lace. They were doing each other's makeup in preparation for the Spring Feria, and were chattering away with excitement and animation.

The young men on the train were taking heavy notice of the preening and primping and I silently watched this gentle dance between a short-haired, young lad and an olive-skinned stunner along the aisle between seats. I arrived a few hours before my man friend and headed to a hostel a south of the center of town. As this was deep into festival time, accommodation was hard to come by and very pricey. Rates were nearly double what they would have been any other time of year, but I was happy to be there.

Seville is a very warm place, and even in early spring the evening air was balmy and probably still around 28 degrees. My back was covered in sweat when I reached the hostel and I had to shower immediately to rinse the day from me. My companion arrived a few hours later and it was a joyous reunion. We hugged and chatted, caught up on the doings and goings of things I had seen and things we wanted to see while we were there. We left the hostel and made our way over to the fairgrounds, catching a bite to eat at a small tienda with barred windows along the way.

Going to a festival in Spain is the best kind of chaos to experience. Thousands of people were scattered all along a vast open field, and stages were set up in every possible space. Mini-platforms and dance halls dotted the concourse, and makeshift bars and pubs, along with eateries of all kinds and every ride you could possibly imagine. Children and adults of all ages were out way past twilight, smoking cigars and drinking wine. No one was overly intoxicated but everyone was laughing and carrying on while leaning far back on chairs with their arms draped along whoever happened to be casually sitting beside them. We met scads of people, both local and foreign.

We went on the Ferris wheel, which took us far overhead of the entire carnival and we caught a glimpse of the magnitude of the festivus we found ourselves in. We cuddled in the evening air and swilled wine. I was invited up on a stage to try to dance the flamenco, alongside several gorgeously curvy, well dressed women with their tiered skirts and elaborate headdresses. I lost my footing and fell off the stage at one point but a short portly man caught me at just the opportune moment. No one seemed to mind so we all just laughed and carried on. Ahh life was splendid, relaxed and full of joy. It was a satisfying feeling. I felt like I was where I was supposed to be, when I was supposed to be. We spent a few more days touring around Seville, eating all the tapas we could possibly eat, wandering along the river bends and taking in the heat. We bought tickets at the famous bullfighting place whose name is very long and well known. They were not the cheapest things, and it was for a bullfight I wasn't even sure I would be able to stomach the following week. It was a Thursday and we decided to go to Portugal for the weekend, to be back to take in the centuries-old practice of dancing around in a pit with a very large, very angry creature. We took off, this time heading west, along the southern coast to the Algarve.

CHAPTER NINE
THE HOT, HOT HEAT

When we arrived at the bus station in Portugal, we were immediately swarmed by the usual dozens of pencione and villa owners and operators. Just like in many other European cities your accommodation sought you out, not the other way around. Most of them held cardboard placards on sticks showing photos of what their accommodation was like, including room size and whether or not meals were included, and you could usually haggle a good price. My British travel companion and I had already picked out somewhere to live out of the trusty travel guide and had started walking along the cobblestone roads along the Oceanside and past a marina.

One woman was rather tenacious in her efforts to have us stay with her. We noticed her in the initial pack that we had darted through when getting off the bus, but she continued to follow us along the road in her blue two door Fiat. She was very nice and polite, probably weighed no more than 100 lbs at the time, was less than five feet tall and had short dark hair. She was wearing blue jean shorts and a cotton blouse. She pulled ahead of us, stopped her car and insisted we stay with her. We reluctantly started looking at the photos and tried to make sense of the usual, rather hodgepodge English translation of amenities. We looked at each other and haggled her down in price. It was about 11 euros a night and was only a block or so from one of the best, most stunning beaches in the Algarve and included breakfast. She insisted on taking my backpack off my back for me, but my male travel companion wasn't having any of that and threw it in the hatch. Eventually we got our luggage and ourselves crammed into the compact vehicle and she threw the car in first gear, tearing up the winding stretch of road along the coast.

The southern coast of Portugal is absolutely beautiful. It is right up there with the stunning colours of the Greek Islands, and has a much more laid back, less touristy feel. The colour of the ocean transitioned from crystal clear, overlaying golden sand, to a deep teal, to the richest azure blue. It is sparklingly clear throughout and the entire coastline is dotted with mini-islands that jet up out of the ocean like Neptune's Trident, marking the way to paradise. The Algarve is hot, and even in early May it was nearly 30 degrees Celsius by 11am. The breeze was the perfect temperature, but for anyone who does not like heat, I would suggest taking in the sights in March or November for more comfortable temperatures.

We arrived at the villa, which was really a three bedroom apartment that had been converted in to a four bedroom makeshift traveller's mini hostel, and were happy to find a group of four others had already been staying there for some time. There was one American girl about my age, one British guy a few years younger than me on his gap year, and two Australians who were also just a few years younger than me. They were amazingly warm and inviting, and we all got along fairly easily right off the bat. The woman who had stalked us into staying gave us some keys, collected a few nights of rent and quickly scooted backwards out the front door with some quick instructions on where to buy groceries and how to let her know if we planned on staying longer than the two nights we had planned on staying originally. My British companion and I made our way back to town near the marina and enjoyed our first dinner at a very lovely restaurant overlooking the hundreds of boats parked on the water. This moment spent leisurely nibbling away while at the marina was one of the last very quiet, tranquil and peaceful times we would have together for the next few weeks.

We made our way back to our accommodation to find our other four flatmates embroiled in some sort of drinking game. All four of them, and all six of us, despite our different heritages and backgrounds were extremely boisterous, happy-go-lucky and basically very nice people who all managed to get each other excited about almost anything. We joined in the game and decided to let our new flat mates show us around the town's nightlife. We ducked into cafes, bars, pubs and outdoor discotheques. Two of the six of us had dreads, which usually ended up being the international symbol of Bob Marley and friendship I quickly learned, so we made friends very easily. We danced for hours to North American top 40 music, mixed with a little Euro electro until the early morning, sweat dripping down our backs because the nights remained rather warm. We didn't need more than a sheet to keep us comfortable once we returned home and spent the following day soaking in the deep rays of the sun at the coastline. We quickly realized we had wanted to stay here in this paradise long past Sunday. We forfeited our tickets to the bullfight back in Seville and booked in for a few more nights at the inn. We managed to make friends with the locals and found some of the best food in the area, which sometimes happened to be locals' houses that had been turned in to restaurants to draw on the influx of traveller income.

Steak on Stone was, to be sure, the number one most often-ordered dish by my male travel companions. I was still not eating meat, but I heartily enjoyed watching the process. The server would first bring out numerous little ramekin dishes with your sides and accoutrements, followed by a choice of a few vegetables and a searing hot brick of stone that would be set out in front of you. Then they would bring a plate of the meat in the cut of your choosing and my friends would decide which part of that cut they would first place on this kiln baked brick that was subtly steaming in front of them. It reminds me a bit of Korean BBQ, without the Asian spices. You cooked your meat to the timing of your liking, added sauces, sides and seasonings ad hoc. It never got old with my friends, and it sure was a fun process to watch. The two Brits often regaled us with tales from England, and the Aussies were not shy in their love for their country. They had raised an Australian flag in the room at the Inn and we regularly took turns making homages to the Queen of England and the stars of the Southern Cross.

After a couple of late nights in town and days spent at the beach, we decided as a group to head off into the Portuguese wilderness for some good old-fashioned camping. We had spoken to locals who told us about fine camping destinations. One of the Dred-headed Australians had made friends with a local man who gladly offered to drive us to various trail markers. We loaded up at the local market and headed out to a narrow sandy trail. We said goodbye to our local chauffeur and walked towards the ocean for a time and set up camp a few metres from the shore. There were a few other tents around, mostly couples out exploring. By and large we were completely by ourselves. We ate simply, drank simply, and completely melted into the sand on our mostly private beach on the southern coast of Portugal.

It was here between midday naps and carefree swims in the ocean that we decided to head off on a much less-anticipated journey further south to see Morocco. A few of our Portugal travel mates were not interested and decided to see what Lisbon and Porto had to offer. The two Australians, my British travel mate and I decided to travel back to Seville to start figuring out what we needed to do, and how we could possibly get across the Strait of Gibraltar. From what my travel guide could tell us, and from what others had passed along in conversation, there were a few ways of going to Morocco, including by ferry to either Tangier or Ceuta or by plane to anywhere. I basically took the lead on coordinating the trip. The Australians were perfectly comfortable doing whatever, whenever, with whomever, and my British companion had a much similar opinion.

From my readings I decided Ceuta would be the easiest as it is still considered a Spanish territory, so we wouldn't actually have to deal with border patrols or immigration until we decided we were ready to cross that line which we all thought would be invisible, like most other country borders. I was very happy to be doing this leg of my journey with three males. I don't know if I would have felt quite so safe or confident if I had been alone, or even with another female. I knew Morocco was going to be my first experience in a Muslim country, a religion I honestly still don't know all that much about. I read up as much as I could about respecting local customs as much as possible. I surely did not want to make myself stand out more than my "hair as straw blonde as the wheat in the fields" did for me. I had decided to keep my scarf close and to wear the only pair of jeans on my legs at all times. None of us really knew what to expect, but we were all very keen on having an adventure.

The ferry pulled up in the port town of Ceuta, a large, steep, rocky city along the eastern tip of the Strait of Gibraltar. So far everything seemed the same as mainland Spain. There were tiendas and restaurants along the bay, and families out walking and chatting. It was getting to be late evening and as I was the only one at the time who could speak any amount of Spanish, it was up to me to ask the locals how we would get to the edge of the city and cross the border. Time was ticking and we were all getting anxious to find somewhere to stay for the night. We had been up relatively early that day in preparation for the ferry ride from Algeciras to Ceuta and were getting tired.

We started hiking through the city towards the mountains. No one paid us much mind, and a few were happy to point us in the right direction towards the border. One minor point I seemed to not pay much mention to while researching this area was that the border that marked the end of Spain and the beginning of Morocco was actually not an invisible line, but a ten foot fence, topped with barbed wire and regularly patrolled by a very large fleet of military. I can't speak for my travel brothers, but this was the first time I saw something quite so massive and far reaching as an actual physical fence dividing land. As we looked into the distance to the west, we could follow the line of barbed wire along the mountain range and in to the sunset. I was fairly anxious at the border crossing, mostly because of the police presence, and the extremely official feeling this facility had in comparison to any other border crossing I had experienced. I wondered who they were trying to keep out, or in for that matter. Most crossings seemed like more of a formality than anything, like a check-in so the country knew who came in and who went out. This felt much more substantial, even though it didn't end up being anything more than a check in. I actually thought for a moment I was going to be thrown into a room and searched numerous times, but most thankfully that did not occur.

Once all of us four had been processed and had our passports stamped, we exited at the south end of the facility and walked across the parking lot. At the time the process that was written about and talked about was that once through the customs station you could walk to a cab stand and continue on, as though you had just stepped out of an airport. The only detail we were not prepared for was just how many cabs and drivers would be at the other end to receive you, and there were hundreds. They were spread out across a large open dirt field, all yelling and shouting, mostly in Arabic and some in French. As I again was apparently the linguist in the group, I headed up the task of speaking French to try to figure out who was the best choice for the transfer into Tetouan, which was about half an hour from the border. It was late, and the cab drivers were now shouting, pulling and pushing all four of us through this melee of cars and people. Some cars seemed more local as they were just throwing their goats and chickens in the trunks of cars and speeding off.

I held the hand of my British mate, probably hard enough to crush some bones but it did make me feel safer than being separated from anyone I knew in this new and seemingly different country. I signalled to one man, who had kind eyes, and a rotund belly. He was wearing a traditional djellaba in off-white with some red and yellow striping down the right hand side. I asked in French if he would keep me and my friends safe on our first night in his native country, and he said he would. I really had little choice but to take his word and we all piled in to his taxi, our four backpacks squashed in to the trunk, and sped off into the hills above, with the sun now fully set into the night.

Miraculously I was able to act as translator during the ride into Tetouan. I'm not terribly sure how clear or how accurate I was being though, but he didn't seem to mind. I also tried hard to throw in the few words of Arabic I had learned during the voyage across the sea. He seemed appreciative and was a very nice man. I pointed to a place we had all agreed we would like to stay in our guidebook, and he whisked us up the hillside going very quickly along very narrow roadways. The taxi was quite old and seemed like it had seen its fair share of people and places. Some of the streets seemed to climb at almost a 50 degree angle, and were scarcely paved. We all stared out the windows as the houses zipped by us in the moonlight.

We arrived at the building that had a small yellow rectangular lit sign with red writing hanging off a metal frame, and climbed what seemed like eight flights of concrete stairs with little to no lighting. The taxi driver came with us and spoke to the operator of the pencione, a small quiet woman and assured us this was a good woman and a good place to rest for the night. We were all a little ragged at that point, starving and incredibly thirsty. My jeans were sticky with sweat, even at 11 pm and my body ached from the stairs. I thanked the taxi driver in English, French and Arabic, paid our fare in euros, and we all made our way out into the night to fill our bellies. There was not much open at that time, so we ate snacks we found at a corner store and nuts we had brought with us from Spain.

The next day, once we had a good night of sleep behind us, we sat and decided our course of action which was to rent a car and head farther south. We had discussed a few cities we would like to see, a few Moroccan experiences that were "must do's" and we walked through town until we found a car rental office. It was very hot outside and not a cloud in the sky. I found out very quickly I was also the only one amongst the four of us that had a driver's license, so I also got to wear the hat of the chauffeur, which came as a bit of a relief as I dreaded getting car sick as a passenger. The car rental office required me, as the only licensed driver, to leave behind either my credit card or my passport. For some very stupid reason I thought it made more sense to leave behind my passport. I thought if I left my credit card behind they could just go on a shopping spree and didn't realize that an official government document had a little more value. I'm not really sure why they needed either document as they already had the imprint of my credit card: apparently that is just how things were done.

We climbed into our new blue four door rental, with hand-crank windows and a functioning stereo and hit the road out of town. The drive to Chefchouen was a rather interesting one for me as I was now navigating all four of us through a country where I had never been, and realized soon into my driving experience that road rules in Morocco were more like subtle and loose suggestions on how to conduct oneself. There were cars everywhere as stoplights were not necessarily adhered to, as were directions for the flow of traffic. There were also herds of sheep and goats that you had to share the road with, including their herders with their long, hand carved staphs fully kitted out in djellabas. They also wore well fitted and well spun shesh or touaregs tightly wrapped around their heads. The animals and cars shared the road symbiotically in most cases. I did not see much in the way of spilled blood of accidental deaths of animals on our travels.

My makeshift headscarf was a loosely knitted purple scarf I had borrowed at the last minute from a friend's girlfriend as I was leaving the United Kingdom. I wore it every day over my braided hair as it worked well to keep my scalp from getting burned and seemed to keep the cat calls to a minimum in most places. It was one small customary thing I could to honor the country I was a visitor in, and it actually felt great to have the ability to experience this country for all of the people, customs and traditions. We spent the day walking around Chefchouen to try and find a djellaba for one of my Australian travel mates as he was dead set on embracing the traditional dress. The town itself is right on the hillside and surrounded by lush greenery. Most buildings were whitewashed and densely packed together. The houses were of similar shape and had simple doorways and windows without glass. Carpets hung out of windows laundry was hung out on rooftops. Every local knew very clearly that we were outsiders and because of the Aussie with dreads, we were called to and referred to as "Rasta Man". We had probably a dozen offers to go visit "my brothers' farm", and could hardly make it down a single street without someone offering to sell us hashish. I was very cautious of this, as our intrepid guide book very clearly warned us about the hills around Chefchouen and all the possible situations we could find ourselves in.

We all laughed it off and continued to speak to the local shopkeepers to find just the right djellaba. My Australian friend was rather tall, and he was looking for certain colours and placements of stripes and things. We finally met a very nice man who assured us he had just the thing, so we went inside his home where he had dozens of this customary dress hanging, all hand-made, in place. My friend did end up purchasing one from this man, in his living room right near the sewing machine that stitched it into being. You can't get much more authentic than that!

It was late afternoon by this time and we were beginning to think of where we would like to stay, or if we would continue on driving, when we were approached by another local. He could not have been more than 30 years old, and continually told my travel companions that his brothers' farm was the best one in town, and only a two minute drive from town, and basically persuaded my group of men that this is where we should go. I'm not going to lie: this felt like a very bad idea to me, probably the worst one yet. I expressed how uncomfortable I was with the idea of driving with this man even if it was only two minutes out of town to visit this world renowned agricultural scene, but my travel partners were set on the idea. So all four of us and our new local guide packed into the rental car and made off in the early evening.

Sure enough the "two minute drive to the countryside" ended up taking two hours and we arrived at a large house otherwise known as a riad. The home was white and beautiful, with archways and intricately-tiled floors. The courtyard was quite big and could probably fit another house inside of it. We had an astounding view of the Moroccan mountains from where we were, with the sun setting warmly over the fields as it was now deep in to the evening. Our hosts were rather gracious in welcoming us into their home and poured for us deliciously sweet, minty tea followed by an absolutely amazing dish of shakshuka, a traditional skillet meal with eggs and vegetables. It had been a long day and I happily filled my belly with the soupy mixture on fresh bread called khobz.

I hadn't felt the safest being out in the mountains, so I kept myself wedged firmly between two of my man friends, and kept one hand on my British travel mate at all times, and somehow this gave me comfort. The men started having conversations with our gracious host in English, but I largely stayed out of it, choosing instead to focus on the room we were in with its very simple banks of seating that made for one large four-walled couch, the tiling on the floor, and the sparse wall hangings. Our host left us for the night and I swore at the boys that if anything happened to me in the night I would hunt them down in the next. But luckily we were fine as we slept like logs and woke up feeling refreshed and ready to go. I absolutely do not in any respect think driving into the mountains of any country is a good idea, and still think this was a bad situation to have been in. I am without question unbelievably thankful that we were kept safe, and that the people we met were intrinsically good people.

Our local host fed us another simple egg breakfast and more tea and we all piled into the car once more. We filled the car with petrol once we were back on the highway and had decided to drive to Fes. There was a hammam along the way that we all thought would be an exciting experience. The locals in Chefchouen had mentioned it and our guidebook spoke to the healing properties of the mud baths. It took a few hours on the largely deserted highways to make it to the hammam. The countryside was mostly farming of grains and animals. We had to stop at an intersection to let a herder with goats, chickens and a few ducks cross, and they all seemed leisurely and carefree.

One thing about time spent in Morocco is that it felt as though time didn't actually exist. The sun rose in the morning, and it set in the evening. Everything in between still happened, and all at its own pace. Of all the hundreds of people who had come to this traditional mud bath house, there were only a half dozen from the western world. The grounds of the facility was laid out for about a city block in every direction and there were big boards and signs all along the walls of the entrance outlining in French and Arabic what types of services that were available. There were people of all ages scattered amongst the rooms, from infants to the elderly, dressed in traditional gear, floating about from pool to pool, and wandering the fields surrounding the baths. There were separate rooms one could rent by the hour that had single bathtubs hooked up to piping that supplied mineral mud straight from the source. The smell of sulphur was strong and I had no idea how to manage any of it so I opted not to sit in a tub of mud this time. We did spend the day walking the grounds and speaking with the other guests. It was a world renowned place that Moroccans themselves went to on special occasions.

We made it to Fes in the evening and found a place to stay that was out of the Medina and therefore vastly cheaper. Our rooms were in a huge building that seemed sparsely populated for I did not see another soul walking around at any point in our few days stay. Like most places in Morocco, the toilets were not the standard western toilets but normal for many countries in that they were squat toilets with two foot mounts and a hole in the ground. I actually had no problem with the idea of using squat toilets as it is a very natural position for me, being all bendy and flexy, and it was much like when I was camping and using nature's bathroom. The only trick with these toilets was the ridge at the front of the hole that is meant to funnel liquid waste into the hole, but I only found it to deflect mine and sent it all over my feet and ankles. It took a bit of finesse and I did eventually get better at it, but not the best. Usually there would be a bucket and a bar of soap nearby for use in cleansing afterwards but I still brought toilet paper with me wherever I went from my days in Cuba and I found it to be helpful. Some of the bathrooms I used I'm not even sure I completely understood how to use the toilets. Some had hoses, some had nozzle type spray guns. I was too embarrassed to ask, but if I ever go back I will be sure to get the entire instructional manual before going in.

The showers in this massive building we were staying in were in the very dimly lit basement. The doors and walls of the shower stalls were made of metal, the walls high and very few windows to let in the light. The water was tepid at best, and came directly from cisterns on the roof. I didn't spend much time basking in its tranquility--it was more to rinse off the day of dust and humidity. We went off in search of the Medina which as a massive walled city within a city wasn't all that hard to find. There are entrance points to Medina all along the outer walls. Outside the Medina it is as city as cities come but inside the Medina is completely different. There is no car traffic as everything is on foot. The alleyways were very narrow all throughout the Medina and the walls jut sharply straight towards the sky on either side of the cobblestone corridors. There are shops everywhere, selling everything you could possibly imagine, and children playing in every empty crag and stairwell. Oddly we didn't really see many tourists while we were there. Our entire time in the country we saw very few which I find strange as it is a most beautiful country. I would most definitely take my family to re-experience and explore further now that I know what an amazing place it is.

Navigating in Medina is tricky if you are not a local. Even with a map or two we found ourselves getting lost at almost every turn. We would end up in dead ends and go around in circles trying to get to the places we were seeking out. We ended up stumbling upon a restaurant in the middle of the walled city that had about four levels of seating where we dined on every variety of tagine you could imagine, presented in the traditional clay cone shaped lids. The food was amazingly flavourful in the Medina, and the cooks clearly knew what they were doing. It wasn't at all hard to find vegetarian dishes either.

The restaurant had a rooftop viewing deck that overlooked the entire city, walled and un-walled. We could see little black plumes of fires billowing up from houses and Mosque towers in every direction. Many of the alleyways were structured so that you could see these elaborately ordained towers from deep within the Medina for prayer times. Many of the locals within the walled city welcomed us with open arms and took us inside their homes and we sat inside the inner courtyards speaking with them in French, learning about their everyday lives. I wore a wedding ring on my ring finger the entire time we were there as I had been speaking with some travellers in Spain who casually let me know that being an unwed single woman travelling with three men may not be the best idea, so I elected my British mate to be my husband. I was surprised at how many people did ask whether or not I was married, and were much more satisfied when I would point to my British husband. Some mothers actually tried to have me meet their sons which I took as a huge compliment.

As we left the Medina on the first day we were approached by a small boy at the exit point who handed us a brochure and told us in English the services he had to offer. At this time there were not many people who spoke English, so this came as a surprise to all of us, but I had read about these lads in my travel guide. He wanted to be our guide through the Medina as he was a local and knew his way around since the day of his birth. The men were all in to hire him for the following day and I tried to have them read the entire section in my travel guide about being hustled by these young prodigal sons but they were having none of it. Sigh.

The next morning we met the young man who was all of maybe twelve who had waited for us at the same entrance point. He absolutely knew where we were, where we were going and just what exactly we needed to see in order to experience all of the Medina at its finest. He also took us to every shop that was owned and operated by his "uncles". One of these was a clothing store that sold every manner of traditional dress for a Moroccan, including rack upon rack of djelabbas. The men went off in search of new robes and I stood and haggled with this new "uncle" over a purple kaftan with white piping stitched along the neck and sleeves. It was very beautiful and simple, with white crocheted buttons up the front and swirls at the wrists. I had to haggle because the price he asked at the beginning was completely outrageous at the equivalent of about 100 British pounds. I had done some shopping around at where the locals shopped and I knew these shirt-length kaftans had a going price of about one pound so I was expecting some markup, but not quite as extravagant and ostentatious as he began. I spent close to an hour trying to whittle him down. Once I got nearer to 30 pounds he became sad and talked about feeding his family and making an honest living. Once we got nearer to 30 euros he became indignant, crossed his arms and refused to make eye contact with me. I handed him 30 euros and he passed me the shirt, and then we both had a good laugh. I think I actually made the grown man in the Medina sweat for a brief moment. I had never thought myself to be a good negotiator until that moment.

The young lad then took us to his other "uncles" who happened to own a rug shop, total coincidence I'm sure. I had read in my guide book about how pushy and forceful in particular the rug trade was, so I had prepared myself for this. I tried educating the men, but as always they were uninterested. They kept insisting they just wanted to experience it all, so experience it we did. The moment we walked into the store which had piles and piles of Berber rugs in every colour, size and shape we were surrounded by six shop stewards who immediately separated our group in different directions, showing us all different piles, colours and shapes. There was an upstairs to this shop as well with even more rugs. I understood the logistics of how exactly I was not going to transport a rug of any size to any place, let alone figuring out a way to get it back to the UK on the plane. I hadn't come to Morocco to acquire things as I was still largely homeless and living out of a backpack. They were not terribly interested in my reasoning that I would not be able to take it anywhere, let alone carry an 80 pound rug through the Medina and strap it to the roof of our rental car. I graciously told them about a hundred times that I would not be buying a carpet that day and went and found my husband, who was having the same sales pitch. His salesman was a little nicer, but looking across the floor I saw my poor Rasta Man friend getting drilled by two salesmen. It seemed as though they were making some headway as he reached for his wallet. I quickly gathered my men, who were red-faced and stunned by what had happened. None of them had any intention of buying a rug, but when faced with a few intimidating people, and separated from their tribe it's more difficult to say no over and over again. We told the boy we were done with his services for the day, paid him his days wage and left the Medina. The following day we laid low and digested our experience from the day before.

We made our way to Meknes and stopped at a fast food restaurant to enjoy some western food and western toilets. One thing I knew for certain was that anyplace in the world with fast food chains always had toilets with seats and toilet paper. It was always interesting to see what local fast food chains had on their menus that incorporated local cuisine. It is all junk food, I am aware, but it was always really interesting to note a falafel burger or a curry chicken burger.

We toured the city and continued on the toll road to Rabat. Along the way we passed through various little roadside villages and pop-up stalls selling necessities like bread and eggs. Many times as we were driving through these villages, the people would gather in front and around the car to see who we were and what we were about. They tried to sell us their wares, holding a goat on a string in their other hand as they spoke and gestured strongly with their free hand. I really liked Rabat as it is an oceanside city with an oceanside feel. The breeze was gentle and kept the air temperature comfortable. There were people lazing along park benches that dotted the boardwalk that ran along the beach.

We checked into a small hotel a few blocks from the water and I ran into the problem of not having a passport for check-in because I left mine with the car rental company back in Tetouan. For whatever reason I thought it was better to give them my passport than my credit card at the car rental shop. I would now never do such a thing, but what did I know then? I had no government identification to give reception. My brilliant British husband figured out a way around this though. I went up to the room and began to unpack and laid my body out on one of the two single beds in our shared room. He went down to the counter and told the woman his wife and he were embroiled in an epic disagreement for which I was now no longer even speaking to my husband and everything was on fire in our relationship. I felt terrible! I hated being part of a lie, even if it meant I had somewhere to stay. It was a nice little hotel and very comfortable. I ended up having to dart around reception during the few days we stayed there due to not having proper identification. I tried my driver's license but that did not suffice.

We went to old ruins and local markets and spent a day at the beach. The waves at the beach were incredible. I wondered why no one was surfing them as they were right-handers and went for miles, with perfect breaks and in perfect sets all day long. I'm not the best surfer, and it is a cause for celebration if I even get to stand and get carried on a wave for a few moments, but it is a fabulously fun and freeing sport. I rolled up my jeans and waded in to the warm Atlantic Ocean. I didn't have the courage to try and get away with slipping into my bathing suit, despite the fact that the beaches were completely deserted. I was happy with having my right foot in the right hand side of the continent's ocean.

Later that day we met some very friendly locals who told us about a pub in town. Alcohol was not widely and openly consumed in Morocco when I was there. This pub was a bit of an underground operation that people only heard of through word of mouth. Obviously the men absolutely wanted to go for a few beers, and again I tried in vain to have them read the warning about being invited to such establishments, and again I was vetoed. We showered up and I donned my new lilac-coloured kaftan out to this pub. It was upstairs in someone's converted living room. There were about a dozen plastic round tables and chairs set out with tablecloths covering the tabletops. There were a few other groups of people and some very timid local couples already sipping on spirits when we arrived. The menu had about three items on it, one of which was small glass bottles of beer that had white labels. We ordered a round of these for us each to have one and our new "friend" also ordered himself one. We started discussing culture, religions, and customs in an open and friendly manner. We had four different countries sitting at our table so it made for some rather interesting comparisons.

A second round was ordered and conversation continued to flow. A few more people came into the pub which had a security guard at the top of the stair entrance and one at the bottom on the street. And then came the third, fourth, fifth rounds and so on. I was absolutely at my limit, and the boys carried on, including our new friend. A few hours later, in the early morning, there were just a few left at the establishment and our friend excused himself from the table to go relieve himself, but never returned. This left us with not only our own bills, but also the bill of this man who was twice the size of any of us, and drank twice as much.

We settled our bills and went to leave when both security guards stopped us at the exit and told us we were not allowed to leave until we paid for this other local man's bill. We all had jaws wide open at this. The men tried to explain we didn't really know this man who had racked up about a hundred dollars' worth of alcohol. They simply did not care and we simply did not have the money. We tried explaining our lack of funds, and again they did not care. They told us they would follow us back to our hotel, and take us to a bank because we were not allowed to go until we paid. So in the dead of night, in the empty streets of Rabat, we walked back to our hotel to get our credit cards with these two very tall, very large men dressed in all-black following us.

We sat together in one room trying to figure out how to handle this mess. We were all pretty scared that we were going to end up in the back of a van in the middle of the desert so we hurriedly packed up all our belongings and devised an escape. We were on the third floor of this building which had a window to the street below, so we began to crawl out the window and along an external ledge. Our thoughts were to get out the building without going through the main lobby where the henchmen were waiting, then jump in the car and drive through the night back to Spain, or to Tunisia or to anywhere that wasn't here, in this moment.

One of my Australian friends had already gone out the window and was slowly making his way scaling the side of the building when I quietly threw my leg over the window opening to make my way out the ledge. My heart was pounding and my brain was dizzy with beer. My fingers tingled and I was actually half laughing in discomfort and half terrified my male travel companions had again gotten me into a situation which I wondered if I would leave alive. I was standing on this ledge with my hands gripping the windowsill, shimmying slowly to the right of the outside of the hotel, when I noticed the security guards exit the front doors of the building and walk away down the road into the dark of night.

Everyone stopped dead in their tracks. I and "Rasta Man" hung out on the ledge, the other two men silently crouching inside the windowsill in the hotel room. Our eyes darted around to each other as we waited for both of them to be far enough along the road to sigh a large relieving breath of air. Little did we know how easily sound travelled in the desert and they both turned around and saw us both hanging there. "Oh my god" I thought to myself, would this be the moment I die? They both had quizzical expressions on their faces, then shrugged their shoulders and let out a half-laugh and continued on their walk into the night. We were all pretty happy and relieved as we made our way back through the window to our hotel room. The interior boys hoisted my sweaty, shaking body back on to the bed and we all laid there laughing and staring up at the ceiling in complete bewilderment.

The next day we decided to make our way back to Europe where we would hopefully not get ourselves in as much trouble. We followed the coastline north, stopping every so often to get something to eat or fill up with petrol. It was a fairly relaxed day as we took our time with the drive but did not have much desire to stop for great lengths. We returned to Tetouan and I was very happy to get my passport back. I had gotten anxious on the drive north thinking of whether or not they would even return it to me. What if I would get stuck in this country? How would I go about getting another passport? Where was the consulate? The moment he handed it back all of that melted away. We took a taxi back to the border and returned to Spain that night. The Australians went back to Portugal shortly after, my British husband went his own way and I travelled the Mediterranean coast to Malaga. Eventually we all met up back in Barcelona and had a very fun time together going out to bars and clubs, restaurants and cafes. We reminisced about our experiences in the south, good for a laugh after the fact.

I remember clear as day the things and stuff that were in the markets in Morocco. I didn't buy a whole lot of stuff, save for the one kaftan shirt I still have. I found it hard to pick up souvenirs and continually acquire things. I had nowhere to put them, nowhere to store them, and my practical mind simply could not justify giving up the precious luggage real estate for something that may or may not sit on a shelf in the home that I did not have at that point. I have my memories, and a few photographs. I have a few items of clothing which despite having travelled countless continents since and might no longer fit, I still hang on to them, as well as my vivid memories. These are my souvenirs. I have a recurring dream about finding this incredible market that I once visited again and buying all the art and artifacts that I had to leave behind throughout my travels. I see clearly the tables of trinkets and hung art. I think one day I will go back to many of these places and revisit the sights, sounds and smells of the things I left behind.

CHAPTER TEN
BRISTOL, THE SECOND

When I arrived back in Bristol after my months spent in the heat of the southern European countries, I was once again homeless, and really had no idea where I fit in. My experiences had left me wanting more but without a proper place to call home I found myself listing like a ship in a storm. I went back to work at the bar I had worked before, as well as a smaller, more old fashioned one whose patrons were mostly the same four people who spent their entire day sitting in front of the television with a pint of Carling. After a few weeks I knew this was not fitting me right anymore and I decided to take a train south to Cornwall to try and find the beach I had visited years earlier on that very fun surfing trip with my friends. I took a weekend off work and decided to spend the Saturday on the train. I really had no idea where I was going or what I would do when I got there. New Polzeath as it turned out was not on the direct rail route which meant I would have had to find my way from the train station to the coach depot and then on to town. People did not have smartphones or ready access to the internet yet, and it seemed like a lot of trouble to handle for one day, so I stayed on the train. It so happened that the train I was on hit the end of its journey in a town called Newquay on the Cornish coast of southern England. I had never been to this town before, and I'm not even sure if I had heard much about it at all in my time throughout England. I consulted the map that every train station has of the surrounding area and started walking towards town.

Newquay is a gem on the southern coast. It is surrounded by coves of soft sandy beaches and became popular as a summertime getaway destination for the British looking to holiday somewhere within their own country. There is a large golf course separating the two main beaches and a few well-known hotels that served as shooting locations for film and television due to their stately image set against a wild sea backdrop.

I found my way to the beach and sat down in the sand. I had no idea why I was here, but somehow gazing out to the ocean made me feel calmer as I struggled with the realization that I needed to find somewhere to stay as there was no train out of town left on that day. I pulled out a travel guide and started reading about the town, but decided to just get up and get moving. I walked up the steep drive back in to town from the beach and noticed a sign out front of a house that advertised itself as a hostel. I walked up to the front door of Original Backpackers Hostel and started to knock. A short ginger haired man answered the door and looked me up and down "You alright then miss?" he asked of me. I stepped inside and he called for the master of the house.

The man who ran the hostel was a few years older than me, very tall and very blonde. I enquired as to whether they had any room for the night, and at first it seemed as though they were completely full. The young ginger-haired man who had first answered the door, upon hearing our discussion, suggested that he would stay on a cot in the kitchen for the night if it meant I would have somewhere to sleep, and the blonde man agreed. I gave them both very large hugs and made my way inside. I explained my story of how I had ended up there that day and had literally no idea what I was doing or where I was or what I was going to do. I ended up staying for the weekend and decided this is where I should be. I made arrangements to be back for the following weekend and returned to Bristol to pack up my things.

I returned with my one purple backpack full of belongings and settled in to my new home a block from the beach. I found work fairly easily at a variety of bars in town. What started out as casual work at one night club turned became full-time once summer was in full swing. Berties was a nightclub and pub that was attached to a hotel that catered to a large crowd of holiday makers. Most nights it was packed with a few thousand people including stag and bachelorette parties. We had celebrity guests come by most weekends in the summer and there were at least five separate bars within the downstairs venue.

Many of my friends who lived with me at the hostel also worked throughout this complex, as well as at other venues in town. There was a large population of Australians and South Africans in the town and a spattering of a few people from all over Europe. I met some of the nicest people of my life while living here, and everyone felt as though we were on one big holiday. Being on one big holiday in your early twenties also meant one big party, and Newquay did not fail in that regard. I worked most afternoons and evenings among a few different venues and would have a very good time tending bar. I tied for highest average sales with another competitive British friend of mine.

Once the dancing and drinking and laser disco show was finished for the night and the bottles had been sent to the recycle, we would all walk down to the beach for a bonfire at 4 am. There would be music, drummers, dancing and sitting around talking about whatever we thought was very important at that time. I had a pretty solid group of friends to work with, and we would all coordinate accessories for special fancy dress evenings and would go to after-parties at the owner's son's house. This house was on its' own mini island that was connected to the mainland by a bridge, a pretty impressive sight. I started working at the VIP bar more and more often in the summer, which meant I met a few celebrities such as models and professional athletes.

One of my most memorable nights was meeting Katie Price, a beautiful petite model who only drank champagne with a drop of Crème de Cassis. She came in with body guards flanking her sides and sat on a barstool as I served her. People had to pay a large sum of money to have VIP status to sit with her in this heavily guarded lounge. She and I chatted while people came up to her asking for autographs and photographs. I asked her how she felt about having her photo taken all the time, and she explained that it wasn't the photos she minded but the unscheduled groping got old after a while.

It was thrilling to be part of that, and it was also a strange spectacle to see people clambering to touch this five-foot tall human being. I also had the whole Rip Curl Surfer crew in one night during a competition tour through town. They were fun boys and we stayed up all night chatting and watching the ocean. I spent at least a few days a week in the ocean, borrowing surfboards and wetsuits when I could get them. The town itself was quite geared towards surf culture and the waves were decent. I ate sand quite a few times, but was exhilarated despite my frustration all the same.

I felt like a local, even though I was not. I made friends easily with everyone in town and have kept many of those friends to this day. Darren was one such friend and one of the many Australian men I met through friends of friends. He was, and still is, an incredible chef. He would come down to the area of the club I would be working in after his shift up in the kitchen in the hotel with our mutual friends and have the same drink every night, a large pint glass with a handle full of bourbon and cola. Most nights it was more bourbon than cola, and he would regale us with his unique brand of gritty Aussie wit. For all his loudness he was actually the largest teddy bear and had a healthy crush on one of my girlfriends. It was cute for me to watch them banter back and forth at every exchange. They have both since married other people and become massive successes in their own fields.

On another occasion one of the DJ's I worked with at the night club also had a radio show during the day on a local station and he needed a favour. Over one weekend they were broadcasting a dating competition and he desperately needed girls to participate. The radio station had set up a stage with a marquee on the lawn in front of the hotel. He was broadcasting me out to the town, trying to find me a date. It made me cringe but I obliged regardless as I thought it would be a hoot to be on the radio. I had met this young French Canadian hockey player who was on holiday in town that weekend and he agreed to go on a prearranged date with me. The radio station sent us off for dinner at a restaurant in town to see if we would hit it off. He was very nice, and we spent most of the evening talking about home and the province of Quebec. He had been playing for a farm team for the NHL and decided to take some time away to re-evaluate. It didn't end up being much except pleasant company. The radio host wanted an update with a super happy ending or some kind of dramatic Shakespearean outcome, but I had not much to report and he seemed a little discouraged at my inability to find a keeper.

One weekend there was a man from up north with a thick accent staying at the hostel. We ended up flirting with each other one night and then had a long romantic chase through town, not entirely sober. We kissed for hours at the beach, our clothes getting soaked and sand getting all in my hair as we lay under the moonlight. He cradled my head with both of his hands and pressed his face against mine. The man could kiss. I wrapped my legs around his as the ocean waves lapped up against us. There were party revellers all along the beach beside campfires but none of them paid us much attention. I could hear seagulls still out, flying overhead as we leapt up and ran through town stopping at park benches and in the golf course to kiss some more. The tennis courts still had their lights on and we kissed beneath their light, spending time wrapping our bodies around each other. It was nice to just kiss, and I had no intention of doing anything other than that with him although he clearly had other plans.

I was a terrible kisser so it was nice to have some practice. I hadn't had much in the way of physical male involvement in the years leading up to then as I was still pretty timid from my time in Montreal and wasn't completely comfortable with having someone touch my body or touching other people's bodies. I was desperately afraid of having a label placed on me that was unflattering so the next day when this guy started bragging to all his friends that he had snogged rather passionately with me I strongly denied it. I didn't like feeling as though someone found an intimate experience with me to be some kind of trophy. I should have just embraced it, but I couldn't. He felt so bad that I wouldn't acknowledge we had had a romantic episode that he left the hostel. I was so uncomfortable being attached to any man, and it was nothing he did or didn't do. He was probably the best kisser I had ever met at that point but I wasn't even ready for that. It was a beautiful night we shared together though. I am thankful for it as it still makes me smile today.

In the middle of the summer word travelled that someone was hosting a massive outdoor party on the other side of town. Everyone I lived with and worked with had heard about it. We all started walking over to this party in the early evening when one of the guys that regularly would come around at the hostel started to go south on a bit too much substance. He declared he was a crocodile and started wriggling around in the dirt and rocks. Everyone thought it was funny, but I found it to be concerning to say the least. Drugs were obscenely easy to get all over town. During the late night beach festivities people would regularly walk around with freezer bags full of pills, rocks and powders. Nothing was hard to obtain, and a lot of people lost a little too much touch with reality that summer. I was always short on sleep living in a room with eleven other people who all kept different schedules so thought it might be better to sit this one out. The next day by about midday I found crocodile man sitting in the living room muttering about the drugs he had to do in order to come down from the drugs he did yesterday. He had lost a tooth somehow the night before. I didn't see him much after that, and I can't imagine his road got any easier. This was the first time I had the experience of being surrounded by a lot of drug use. It would carry forward into my future time as escapism is fairly common. I had never directly known someone who was an addict until this point, nor did I yet understand the depth to which substances can affect a human being and everyone around that person.

By and large everyone was lost that summer. We all got along so easily because we all had that in common. Most of us hadn't figured out our careers or an educational path to follow. We had come from all different corners of the world, and we brought our emotional strife with us. Many had left their homes in hopes of "finding themselves", and maybe some of them did, but it would have been hard with the lifestyles we kept at the time. I made many incredible friends while here, one of whom was a Swedish-Ethiopian young woman named Emma. She was a tall, dread-headed cook who worked at various local restaurants and pubs. She carried burn marks on her arms and hands from times spent too close to the heat in kitchens and wore them with pride.

Emma had a very kind nature and an extremely loving disposition despite her grand stature and stoic body language. She was my best friend through all the days I spent in this coastal town. She came with me to parties, we grocery shopped together and she also would hug me when I cried. Her abrupt and direct personality is what I admired most about her as she did not take much flak from anyone and would tell you exactly what she felt about any given topic at any time, whether you wanted to hear it or not. It is my hope to one day meet with her and many of those I met in Newquay again, to recount stories and catch up with who they have become. I am thankful I was living where I was, with the group of misfits we all were.

Although we were all meandering we all at least had each other and this one home in common. I believe it has shut down now which saddens me as it served as a bit of a beacon. The owners of the hostel were family, a group of siblings that all shared something in common. Their best friend and brother had been washed out to sea by a rogue wave from the rocks about a year before I had met them and they were all still reeling from the loss. I always appreciated that they welcomed me and a whole host of lost souls. They were accommodating and accepting of all of us as accommodation was scarce in town, they could have easily sent most of us packing. Sometimes I wonder if they didn't open the hostel and keep it going in the off chance their friend and brother would someday find his way back to them, in some other plane of existence. They told the story of the night he disappeared more than one time, and every time I would cry and so would they. They had made a papier mache badger head for his wake that served as the hostel mascot. The VIP lounge where I worked let them have the bar to celebrate his life on the year anniversary of his disappearance. We drank and we cried. I hadn't known him, but I was affected by his loss all the same.

In the fall my girlfriend Janice from Canada, who had been living in the USA at the time, came to visit me. I had made arrangements for her to stay with me and I brought her to work with me on the nights that I worked. I had picked up a few nights working at a club on the other side of town that did a throwback disco night once the summer rush of tourists had left for home and she came along to experience that. It was fancy dress, and we all wore bell bottoms and tie-die shirts. Like most clubs in Newquay this one was also below a bar and usually had great attendance for the fancy dress disco party night. It was a fun time for everyone and I introduced my girlfriend to all the owners and all my local friends and they made sure she had a great time. They played classic soul and funk until the wee hours of the night.

It was so comforting to have someone from home to be with me for a short time while I transitioned. She stayed with me for my final week in England and we both took the train together to London, with her flying back to North America and me to Australia. She has been with me through thick and thin, always lending support when I have needed it most. I have hopefully been able to show her my appreciation for the many hours I have spent weeping less than silently on her shoulder before and since that visit. I was so fortunate to be able to share my last few days in England with her. Many of my friends that I lived and worked with in Newquay were from Melbourne and they had invited me down with open doors at any time to visit their fair land. I happened to have had a crush on a young Australian man with dyed blonde hair and blue eyes, who looked amazing with his shirt off. That, coupled with the realization that it was now November in England and beginning to get very cold, I thought it the right time to fly south and follow the heat.

I have remained friends with a few of the people I met in Newquay, and many of them live in other cities and countries at this point. Some went back to their native lands, others moved on to still newer locations. I absolutely did not find myself in this coastal town, but I did learn how to have fun again. I learned there were still wonderful people in the world, people that did not want to hurt me, and people who were just as lost as I was. I have always been very tough on myself for feeling lost. I thought for the longest time that once I had achieved certain milestones I would not feel so empty. I thought if I was with the right person, or had the right job, or had a shiny car, a fancy house or a thriving business that I wouldn't feel so lost. I thought if I just kept travelling that eventually things would make sense. But none of that held true. As much as I knew I was not working on satisfying myself while living in this place, I felt it was necessary to live this style of life as it was truly burden free. I simply had to live my life, however that was to go, and to survive the choices I had made.

I have since learned that many people feel very lost, just as many people are also hard on themselves for this feeling. I have met people well into their elderly years who can still say they do not know who they are or what they want to do when they grow up. We are all very good at pretending to know what we want or who we want to become, but it is not until we realize that we are continually evolving that we make waves for change and cease being so hard on ourselves for realizing we are not who or where we intended to be. The moment I decided to embrace myself in the moment for who I am and who I am not, that feeling subsided somewhat. I have learned to accept the concepts I grew up with as being goals may not actually be my goals. And when I started writing what I actually wanted for myself, things started falling in to place. But this would not happen for years after my England experience. I still had a lot of learning to do, and a tonne of painful mistakes to make.

CHAPTER ELEVEN
MY FIRST HUSBAND

I've been thinking about how to describe the relationship I had with my first husband. I call him my "first", because he is the first person I married in my life. Hopefully he won't be the last, but maybe not in the same sense as a traditional marriage. I don't think that I would like to have myself legally bound to someone again. I would like to have a "wedding," otherwise known as a big party, with all my closest and loveliest friends and family, but I don't think I'd "register" my relationship with any one god or church, or even government.

I like the thought of a union that is blessed and whole. I would like for religion to be part of it. I know exactly what it will look and feel like, and what the food will taste like. I have visualized it all. And a lot of that came out of what my first and subsequent experiences have been like. I now know clearly what it is that is important and what I want and need in another person, including how I would like that union to feel. It has been a long enough time that I no longer feel any strong feeling either way towards my first husband.

My first summer in Australia was very hot. I first flew in to Sydney and spent a few days taking in the Opera House, the harbours, the restaurants and the shopping. I had met up with a very lovely girlfriend I had met in Newquay and her Scottish friend who was equally as gorgeous. We took a few days to drive south to the state of Victoria, stopping at every picturesque white-sanded beach we saw. When I first arrived in Melbourne I spent a few months living with a family of a good-looking surfer that I was friends with from my time in Newquay.

His family had been most gracious in asking me to stay with them until I figured out what my next destination would be. They had a comfortable house on a large lot in a rural area about an hour south of the city. The first night I stayed with them we went to the local pub to celebrate my friend's return to Australia. All of his friends from high school had gathered in a tiny pub that had a back room with a dance floor. I was introduced to very many people, most of whom had known this blonde-tipped, blue-eyed young man for years. He introduced me to his best lady friend with some awkward words "ah, so I met this girl in England and asked her to come to see me, and here she is!" he said as he quickly pushed me towards her, making a dive for another beer at the bar.

She was gracious about being tasked with showing me around and asked me all kinds of questions to try and help me to feel at ease. We drank beer and premixed bourbon and cola in cans well into the early morning. Bryan Adams' Summer of 69 came blaring over the speakers and I turned to my new female friend and asked why I was hearing vintage Canadiana in a tiny pub in the middle of nowhere Australia. This song is apparently hugely popular down under, and virtually everyone knew the lyrics. I found that shocking for so many reasons, but was happy all the same to be greeted by a piece of home my first night in my new town.

It was November and average daily temperature was in the early 30's. Grass didn't grow and dust was everywhere. I spent most of my days relaxing at the ocean or visiting with friends of friends. We went to the coast for surfing and even further down the peninsula to party.

One night we went to a rather lavish bar that was full of professional Australian Footy players. All my new friends kept pushing me to meet them, so as I normally do I just walked up and started a conversation. All my new friends were staring from afar, as though I was speaking to idols and gods. I had a pimple on my nose at the time, and one of the players (whose name I never knew, but apparently he was kind of a big deal) kept commenting about it as though it was the only factor for why I was not privy to be occupying the same air as he did to the point where I became pretty embarrassed and felt my self-esteem plummeting faster than the Titanic. I quickly learned that this kind of belittling was a common practice in Australia. Men were none too nice to me for the entirety of my stay, to put me in my place and understand that as a "Sheila" otherwise known as a female, I absolutely came second. I am sure not every man on the continent was like that, although that was my experience.

I went with my Aussie surf dude-bro and his friends on a surf trip up north to Forester NSW in the New Year. Instead of breaking up the 14 hour drive with some sightseeing along the way, the surf dude-bro insisted on driving straight through with barely enough stops to use the washroom or pick up some Hungry Jacks from a drive-thru window. He was in quite the rush to get his surfboard in the water and had little patience for anything else. We stayed at a gorgeous campsite that was all sand. The ocean was separated from the campsites by a tall sand dune that served to shelter us from onshore winds. This was the first time I met a goanna, which I actually thought was some kind of Asian dragon. They were roughly half my size and looked like crocodiles that clung vertically to the sides of trees. I would stare at them in awe, but no one else seemed too fussed to see them. Most of the campers regarded them as pests as opposed to gorgeous reptilian creatures.

We spent all day at the beach, swimming, surfing, and learning about just how strong riptides in the area were. I ridiculously coated my body with baby oil in hopes of getting that "Australian glow". Now more than ten years and dozens of new moles later I have learned that was probably not a good thing to have done to my skin, considering the lack of ozone layer around the South Pole. Evenings were spent around a very carefully placed camping stove eating snags (hot dogs) on white bread and sipping cases of Victoria Bitter and Tooheys beer. One night we strolled down to the beach to watch the men surf and right at sunset a pod of dolphins came out to play as they cut through the waves. One of my friends thought it was a group of sharks and started yelling, but quickly realized they were much too playful and understood their true nature. It was a group of stunning grey dolphins, jumping in the waves with the sunset as a backdrop. It was the most gorgeous memory of my entire life down in Australia.

Upon returning to the state of Victoria, I knew it was time to move to the city and get a job to start financing a life of my own. It was fall and jobs were scarce. I went to numerous interviews and trial shifts, one of which was at a café inside a cinema in St. Kilda. I got along well with all the staff and tried to learn the local cocktails as quickly as I could. I was frothing milk in an espresso machine for a customer when the owner pulled me aside and told me that I could not work in her establishment because the sound the froth was making was not nearly the right sound. She thanked me for my time and literally kicked me out as fast as she could. I was totally horrified and aghast that I had been denied a bartending job due to my lack of milk frothing skills so I went to the bar below it to have one friendly drink with the barman, who happened to be friends with the owner. He let me know it was nothing personal and that she was very particular in an attempt to soften the blow. He was very kind and had me drinking for free until 4 am with him. He even offered to give me some heroin for my troubles, but I graciously declined.

I walked home to my new apartment in Elwood from St. Kilda at 4 in the morning, sobbing basically uncontrollably the whole way there. I was not in any way sober, and had just been coldly rejected from a job I expected I would have excelled at. I went for other interviews but was having little luck. I thought my whole purpose of trying to make a go of living in Melbourne was to spend a few months making some cash down south and then slowly work my way north towards the Whitsundays, Indonesia, Polynesia and beyond. However I seemed to get my wheels stuck in the sand in St.

Kilda and ended up on a very unpredictable roller coaster ride that took me years to get off.

My first husband Greg was rather unromantic. He never proposed to me, he never asked for my hand, and he didn't discuss it with anyone. It was actually more of a business proposal. There was no romance and no unrequited chasing, so it felt very forced, and contrived. Everything about it seemed wrong, right from the beginning. I did have feelings for this man at the beginning, but not nearly the rip-roaring lusty feelings I have felt for others. He was handsome at the time, and had an odd sense of humour that wasn't quite like mine, but it existed nonetheless. His family was very generous and took me in even in the very early days and fed me the most delicious food.

Before we were dating he was my boss at a nightclub. The first time I met him I was wearing tight-fitted white pants I had bought at a Zara in Barcelona, and the same grey suede Vans sneakers with a red logo streaked down the side that I had worn every day for over a year at that point. I had on a low cut denim halter top and my hair had been straightened and left loose. The evening was hot, and so humid. The sun glared brightly off the shop windows as I dismounted the tram and headed towards the venue on Fitzroy Street. He showed me around and asked the typical questions of a prospective employee.

I asked what the most popular drinks were and the type of crowd that was here. He replied "mostly Wogs". And I had no idea what or who a "Wog" was, but I learned quickly. I was placed behind a bar upstairs with a gorgeous Middle Eastern woman who shared the same birth name as me, but went by another. She was curvy as the day is long, sassy, vibrant and brimming with life. By surveying the crowd, and hearing the music, I quickly learned this was Greek night. I was actually familiar with a lot of the popular Greek music, due to my time living in Athens and the Cyclades, and I certainly knew the dance moves, the drinks, the slang, the straightforward maleness. Although I have to say I'd not been covered in Ouzo and Cola to quite that extent until then.

It was great, and after months of pounding the pavement looking tirelessly for a job in the "offseason" subsequent to Australia Day, I was happy to be making some money. I and my new boss spent a fair bit of time together for the first few months as we got along easily. I had moved with my roommates--the beautiful Scottish woman I had previously met in Sydney, a lanky awkward Kiwi with the softest temperament, and another fiery, feisty friend from the United Kingdom--to the suburb of Prahran into a gorgeous three bedroom apartment that later would also house my Ethiopian-Swedish friend Emma. When we first moved to this beautiful new apartment I had no bed or mattress and I was sleeping on the floor with my sleeping bag. My then manager friend decided to sell one of his mother's mattresses to me without even asking her when he heard that I was sleeping on the floor. I shake my head looking back for his mother who had little say in the reorganization of her things, but it was a nice gesture. He eventually made money off it by selling it to my Kiwi flatmate when I moved out of the apartment and in with Greg.

He was living with his parents, which was fairly typical in Australia even well into someone's 20s. Melbourne is not the least expensive housing market I have come across and many chose to wait until well past schooling was completed in order to afford a dwelling of their own. I remember the first time I'd spent the night and borrowed his silken boxers with a cartoon character print and t-shirt to sleep in. It was so awkward that I slept on the edge of the bed the whole time, with his parents asleep in the room above us and his sister down the hall.

It was close to 5 am when we got home from work, as nightclubs have 24 hour liquor licenses and we had worked late. The sun was streaming in the windows and the birds were already singing their songs. I decided to walk downstairs to the kitchen to have something to eat and found his mother to be already a few hours into preparing food. His mom was welcoming and explained what she was doing in preparation for giving alms to the monks at the Monastery. Countless curries, chopped salads, fresh poppadum's and baked treats were all slowly being created and wrapped for the journey. The food was very good and she made for easy company even though I felt completely out of place.

Greg and I weren't dating at that point, and we weren't even kissing yet. We went along as friends for a few months working and hanging out together. It was nice to have a friend I could share the majority of my time with. One late night at around 4:30 am we were sitting at Chapelli's on Chapel St. which happened to be a very well-known restaurant that was open extended hours after work. I was eating my most favorite vegetarian ravioli with freshly made tomato sauce filled with olives, herbs and onions and the sun began to stream through the windows. I remarked to him how delightful it was to share in his company and that seemed to be the sign he was looking for to begin to pursue something more than friendship as he was not shy in his intentions after that. I found him rushing me much faster than I felt comfortable with into a physical relationship and eventual coupling. I was not ready for it in the least, but I also felt too uncomfortable to oppose. I actually remember trying to get out of it at one point because it was all too much, too quickly and I was not ready to commit to someone.

When I opened my mouth to oppose or explain tearfully that I just was not ready, I choked on the words and just thought maybe if I gave it more time it would be better. He was treating me well at that time, he sucked the tears off my face like a vacuum when I would cry, and he accepted me for my stray chin hairs and quelled my insecurities when they arose. He treated me to meals and bought me a sterling silver ring I had noted that I liked at a kiosk in the mall. It felt nice to be wanted and accepted and to have the attention of someone seemingly so loving and doting, so it was not the hardest relationship to move through time with. I did have feelings for him, but I felt as though I wasn't given the opportunity for those feelings to grow organically. I predominantly was the one who usually pursued, so to be the one being chased seemed decadent and must have meant true respect, admiration and adoration. Right?

Most of our lives were lived at night, and usually with an unhealthy dose of alcohol present. I had jobs at a few different night clubs in town and was getting tired of living nocturnally and never seeing the daylight. I was getting bored with clubs and venues and found it hard to feel connected with people that were always out for a party. I also spent time doing odd jobs such as a very short stint working at a Kosher Butcher. I learned in three days more than I ever thought possible about brisket, industrial saws and how impatient customers became during Passover.

I tried to maintain friendships with my friends who didn't always work nights but it quickly became mostly impossible. Greg and I had discussed changing to day jobs and pursuing "real" careers a few times. I felt it was on the horizon for me to get into a cleaner, clearer routine. In an effort to freshen up the atmosphere, he took me for a holiday down the Great Ocean Road which was something I had wanted to do since I had arrived but hadn't yet had occasion to get there. We left late in the day, which was a bad idea, but as he was not an early riser it was really the only way. The road west was not a straight shot and I get very car sick on the best of days. I had taken a diphenhydramine to help ease the stomach churn as the car followed the ocean along the highway.

The sun had set long before we reached Anglesea which is a small coastal town with a population of around 2400. Any venue that served food had closed hours before our arrival, but we walked into one place that was still open but was no longer serving food. He had not wanted to stop for food along the way and I was ten levels of hungry by the time we arrived at our accommodation for the night. I had to eat cashews I had found in the car and we shared a bottle of red wine for dinner to get through the night. We continued our drive, for which I begged to drive in order to not feel so motion sick and he obliged and slept for most of the next day.

When we arrived in Apollo bay we went straight to a grocers in order to not suffer the same fate as our last stay and stocked up on eggs and bread. We went to bed around 1 am and had asked for a late checkout, as Greg had wanted to try to hard boil eggs with kettle water. This process decidedly took him well past our welcomed stay and the innkeepers barged their way in to the suite when they could not be patient any longer. I was embarrassed that he insisted on trying to boil these eggs in a mug with water from a kettle. They almost charged us an extra night, for an egg. We finally reached the Twelve Apostles which is a National Park and a Cliffside beach paradise. There are massive chunks of rock that have slowly eroded off the coastline to create magnificent scenery that people flock to enjoy. Set against the powder white and Ceylon blue of the ocean it made for an impressive scene. I was still angry about the whole egg incident as he tried to have photographs taken of us together along the seaside boardwalk. I stubbornly did not want to participate in any of his public displays of affection after having to apologize repeatedly for his behavior at the Inn. He was less than impressed by this and we spent the majority of the drive home in complete silence.

My family had arranged a reunion in Canada during that summer which I also parlayed into a journey back through England, Montreal and various other countries. I spent time with my girlfriend and man-friend in London, and with my family in Montreal and Vancouver. I spent a few weeks going to weddings and catching up with friends in my hometown when a huge wildfire broke out due to a rogue lightning strike. My family and I had walked down to a beach access to watch the flames burn through countless acres of forest during the first few days. At the time we had thought it was impossibly far away, but it quickly gained momentum and size in the breezy arid atmosphere of the valley.

Within days it scorched half of the town and forced everyone to evacuate at a moment's notice. We stood in the driveway of our family home and watched the blaze at nightfall as it quickly crept along the mountain ridge, setting alight one house at a time. Every house looked like a matchstick head the moment the flames licked the roofs. Gone, just gone, entire neighborhoods completely vanished within minutes. Cars were clogging up the roadways and there were police on every corner.

I called my friend's father as my friend and her mother had gone to the theatre. I told him to get out as he was waiting for them to return before he fled with a vehicle full of valuables. If he had waited, it would have been too late. He told me he could see the fire approaching their house through the orchard that backed onto their property. The roads were already blocked off to one direction traffic only, that which would get people out. The air was thick as creamed soup with smoke. Red embers and ashes fell like snowflakes. I drove with my friends to try and save their pets and my family grew impatient with my escapades. They had already loaded up the truck and I had five minutes to get there or they were leaving for the coast without me. I went back and threw my bag in the truck and headed off, moving at a snail's pace with thousands of others seeking refuge from the smoke. The fire never did reach the family home, but a huge chunk of town was completely levelled. Many of my friends lost their homes and all their possessions. Oddly enough, after the fire home construction went crazy, as there was now all of this previously forested land that had been cleared for free and opened up new views of the lake below. I am always training myself to see the positive.

I had made some money from working as a cherry picker for a short period and had decided to fly back to Australia seeing as everything at home was on fire anyways. While I was gone Greg had found us a sweet little one bedroom apartment to rent in an area called South Yarra just off Chapel Street which is a super hip and all-too-happening suburb. It was an amazing thing to come back to, although I'd had no say in the decision. It felt homey and I felt welcomed. The apartment was painted mostly white and had a galley kitchen with a black and white checkered linoleum floor. The living room was small and had wooden plank flooring, the windows were old and single-paned and the sun streamed through them all day long.

I had found temporary administration work for large firms and Greg had finally gone back to sales after much urging. My first administrative position was for a life insurance company in one of the largest business towers on the South Bank. I had not slept well the night before, worrying if I was going to be able to do the job of sorting and filing in a library of accounts. I strolled into the office confidently wearing a blue and white pinstripe suit, complete with crisp white dress shirt. Everyone in the office was welcoming and warm. They did not give me much direction except to point at a stack of papers and an alphabetical divider. By about 10 am I had sustained my first office injury which was a papercut to my finger. I left the sanctity of the file library and sought out a Band-Aid, which somehow got the attention of every staff member from each cubicle in the entire office. I had men surveying the cut, and others looking for bandages, alcoholic wipes and tape. I quickly became light headed at the attention I was receiving for something as minor as a papercut and I proceeded to faint onto the carpeted floor. Awesome, I thought to myself as I came back to consciousness. What an absolutely fantastic first impression. The office receptionist kept pushing for me to go to the hospital, but after clumsily and embarrassingly explaining it was probably due to me being on my menses, coupled with 30 coworkers huddling over me, I just needed some air. My temp firm called me later that day, and I felt horrible as they had obviously heard what had happened. Despite that I did show up the next day and finished the remainder of the contract. I was probably the worst filer-paper sorter they had ever had but for $18 per hour I was happy to be terrible at something.

My next few jobs went smoother. I had one position as a receptionist for CSIRO which was a national scientific organization and a position I found thrilling as it had me in direct contact with scientists, labs and lab coats. I had other jobs working for companies like Ford Performance Vehicles and major multinational accounting firms. I spent a few weeks working for Domaine Chandon at their estate winery as a receptionist. One weekend they were hosting a weekend event sipping champagne and eating all kinds of little appetizers. Given my experience I offered to work the weekend tending bar and in return they gave me a coupon for a free balloon ride through the cloud tops over the Yarra Valley. It was a gift I was unbelievably excited to receive however Greg had less than zero interest and I was not able to ever have experienced such an event. Maybe one day. I also went back to bartending on weekends to supplement my income including one swank, architecturally stunning building south of the city in the Docklands called Cargo.

I met some of the loveliest creatures at this bar. One was a striking Irishman with a heavy accent and the most gorgeous twinkling eyes I had seen in some time. I developed a slight working relationship crush on this man who happened to be fit as a fiddle and knew how to wear leather wrist cuffs just the right way to make all the ladies swoon. I did not ever meet a woman that was able to resist his charm while I tended bar with him. We had our fair share of managers at this particular venue, one of whom happened to be a slim South African with poorly controlled diabetes and a slight drug problem. One afternoon I came on shift to find him crashed out and unconscious on the couch unable to move, so I fetched him some syrup from the soda gun for him and placed it inside his lips. Once he revived he let me know he'd forgotten to eat that day. I urged him to get medical attention but he would have none of it. Things were looking up, we were both entrenched in our work and spent some evenings strolling along the river banks and taking in nightlife together as off-time as opposed to working hours. He was also not drinking quite as heavily, and life was more or less settled.

Neither Greg nor I wanted me to go when my working holiday visa ended, so we discussed getting married. He did not want to move to Canada, and thought that getting married was the easiest way. I remember the moment I called my parents back in Canada to give them the news of our impending union. I was standing on a tour boat along the Yarra River serving for a reception as a weekend gig for a Greek family who hosted party tours on the water.

My parents were less than impressed and seemed rather disappointed in their tone when hearing this "interesting life choice" I was making. They had never met this man, they had never been to Australia and the travel time and distance made it basically unreachable for them to visit. The distance served as a bit of a relief for me at the time, as I wasn't feeling the desire to be judged or to have my reasoning questioned at that point. Everything was zooming along nicely and I thought I had finally found a man who wanted to build a life and have a family with me. He had constructed this whole backstory about how it was all going to happen and it all sounded great.

Our wedding was small, catered by my Swedish friend, on the rooftop garden above our flat. I had wanted all along to be married on the beach in the sand, no shoes and white lilies scattered along my path. But that was vetoed at the last minute because he "couldn't be bothered" to go all that way, which was only 20 minutes by car. I panicked significantly a few moments before the ceremony happened. People were standing all around me as I stood in the doorway to climb the stairs to the rooftop. My head was spinning and I had this moment of clarity where I was just asking myself what it was that I was doing. I hadn't thought this through, but it was too late to do anything about it as all eight of our friends were there.

I had a dress and I had jewellery that had been carefully created by a female friend who was a designer at the time. They were gorgeously created stone and cut crystal earrings, necklace and bracelet. They were dramatic yet clean and she had done a fabulous job at making them. I felt honored she allowed me to borrow them for the occasion. There were all these expectations about me that were rapidly crushing me. I couldn't be the coward and walk away so I took a deep breath and just jumped.

I had a hard time choking the words of our vows out, even when our non-denominational officiant repeated the lines for my benefit. I was crying, and honestly found myself in a surreal cloud of existence with a thousand thoughts racing through my mind, yet I was void of thought at the same time. I had flashes of past, present and future. I felt extreme exhilaration and fear. In the end we were all crying these lavish tears of joy, all over my cotton halter top sundress-wedding dress that I had found a week prior in a small shop a few blocks from my apartment. I kept the dress, not because it reminds me of a most glorious day, but because it did signify a new chapter in my life. It was not what I'd wanted in any sense. I had always dreamed of a beach wedding, of a big party, of all my family and friends in attendance. But I only realized just how disappointed I was after the fact. I had spent so much conscious time and energy trying to tell myself that the way it happened was truly what I must have had wanted growing up, because this is how it happened. Only it wasn't. It was what he had wanted. He had constructed the whole thing as he'd liked, with his stories and plotlines and venues, and I hadn't even noticed how contrived it was until afterwards. I was deeply in love with him for a time and I did believe he loved me as well, in his way.

We took a trip back to Canada to visit with my family and friends so they could meet him for the first time. We had a small party at my parents' house for everyone to meet and greet. On the way back to Australia we spent a week in Taiwan visiting my great friend Mark that I had known since childhood. He had been teaching English in a city in the middle of the country. He showed us around and had us stay with him in his apartment. We ate non-stop some of the most delicious food I have ever had, and it all was prepared fresh in front of us in street carts or kiosks. We rented scooters and visited the local parks, sanctuaries and temples. He took us to have our scalps massaged in fancy hair salons that specialized in cranial massage. We spent our nights wandering through open air street markets and looking over piles of goods from China and other Asian countries that were offered for next to nothing. There were knock-off of every perfume, handbag, toy and clothing item one could imagine. The streets were simply brimming with local life. I bought a few bottles of perfume that had been "factory sealed" I was assured and smelled exactly like the real thing.

We returned to Australia and continued on our newly-married way. Once we were unpacked and settled I opened the bottles of perfume to find they were nothing like the designer scent. It didn't come as a shock, but it was disappointing. Shortly after our union we started spending more time out and about, him more so than me. He just had to be at this club or that one and he just had to be seen with this person or that person. His temper became short and his patience was non-existent. He was no longer doting on me with unrequited love and affection, nor did he even feign curiosity at anything I cared to be interested in. The inebriated blow-outs started to happen more and more often, and usually as much in public as privately.

His cousin once confronted me on a night we had all gone out after a family occasion and told me I was worthy of so much more, and I knew it but couldn't feel it. I cried all over him as he told me the truth in the lies. My husband and I would spend a day or two apart, and it seemed as though he hardly noticed. I started working more to fill my time. I worked briefly for a very well-known bar run by a very well-known and wealthy family that owned a host of venues across town. I had only worked at this venue for a few weeks when I had a few disagreements with the manager. Greg had come to pick me up at the tail end of one of my shifts and he had given me some kisses during the time I spent cleaning up and closing down one private area of the club. They had cameras all over every square inch and the day following they pulled me aside and accosted me for kissing him during my shift. They thought it to have been inappropriate and they expressed their disdain that I could be thought of as married and not ready to mingle with them or their clientele.

The next shift I worked after that confrontation I was working behind the main bar and sipping on lime water when I began to feel dizzy. I went back to a stock area to sit on a chair to see if it would pass, however my legs began to feel weak and the dizziness got worse. I stood up but slumped on the edge of a keg and was having trouble speaking. One of the busser boys came up to see if I was okay, and I motioned to him to hand me his cell phone so I could message my husband to come pick me up immediately. I knew this feeling from the prior experience in England and I needed to get out of there as fast as I could. My husband arrived and had to physically carry me out the front door. Someone had drugged me, and it was enough for me to stop working for that company. I didn't return to work at that venue again nor did I go back to pick up my last paycheque. I can't be certain about who did it, but it hardly matters.

We had started the immigration process around the same time as my visa was expiring. We walked into the centre and his parents had offered to financially help with paying for the paperwork which I was extremely surprised and grateful for. We stood in line to get a paper number from a dispenser and sat with several other non-Australians waiting for our chance to submit the first bundle of paperwork. It was exciting, but I also couldn't help shake the feeling of being purchased somehow like an item in a grocery store. Over the course of the first few years we had to continually meet with our case manager and plead our love for each other. We had to supply photographs and correspondence from each side of the family, and I found it to be an unbelievably stressful time in a little hot room with just the three of us over and over again. The stress of having to prove myself every day of my existence in Australia in my work and home life as well as during harsh scrutiny by someone who worked for the government and had no idea who I was usually caused me to spend the entirety of the meetings bawling uncontrollably. I felt it to be degrading to have to somehow prove my love for my husband and my new country over and over again, especially considering I was not feeling loved nor accepted. I am sure it is not so stressful for others, but for me it was harrowing. Everything just seemed so hard, such a monumental uphill battle. I often thought this was what life was supposed to be like, a constant struggle just to breathe and be accepted. I felt constantly exhausted and found it hard to cope at the best of times, let alone when some government worker was questioning why I did anything the way I had.

And then the bar happened. A family friend had approached Greg with the opportunity to help open a venue in the downtown or Central Business District in Melbourne. I fought as hard as I could against it as I knew it would be so unbelievably enabling. I had only months previously successfully had us both working mostly daylight hours in daytime jobs, like "normal" people. We had lives, we had hobbies, and then we had the bar. But I was determined to make the best of it. I would support him and this business venture and throw myself into whatever needed to be done. It was a goal he had wanted for himself, and I wanted to be a part of something with him that I could be proud of.

I felt like I had succeeded to some extent as I physically hand-stitched the plush brocade fabric for the furniture to custom sizes and shapes. I picked all the colours and I hung all the drapes. I designed the uniforms and sourced the materials. I created the menus once the men had decided which would be the "signature" drinks. The group of men who were involved were very involved with the planning and creation of this new place. His dearest friend painted by hand the logo on the window, and he did a fine job.

Everything just came together so beautifully from the custom cut and stained wood to the mini-library we created near the front door and the hand-made centrepieces on the tables. Together we made this amazing place that was so full of promise, and it did rather well. Business grew and word spread. We had lineups and we became a bit of a "scene". People enjoyed coming to our place to relax and unwind after a long day of work. I met some beautiful people from my experience there and I made friends with local business owners and employees. I soon found myself back deep into the night scene, dealing with DJs and the who's who of whomever.

I dealt with the largest egos I have ever known with deep pocket books and deeper senses of entitlement. I'd watch as Fortune 500 hotshots dropped absurd amounts of money on vintage years of scotch and bottles of rare wine. Or hey, there's Johnny with the tight-collared shirt and $400 jeans who had a lot of cocaine and ecstasy and thinks he's king of the world. So he's going to buy 30 scotch and cokes for his friends that he doesn't really have followed by a meaningless argument with some other patron causing me to step in by crawling over the bar with my soda gun to literally cool them all down before they set each other on fire and tear each other's faces off until security could make their way through the crowd to help them on their way.

One evening we hosted a birthday party for a young woman who was turning 21. Birthday parties for someone's twenty-first year are incredibly important and lavish occasions, larger even than the 15th birthday in Cuba. Families spend tens of thousands of dollars on out-partying their friends. They have specialty cakes made and dresses from designers shipped in. All extended family and friends are invited to participate in the partying, in the photographed slide shows, in story-telling and in drinking. This specific evening saw about 400 of the birthday girls "closest" friends and family members at our venue. They had done the slideshow and told the classic embarrassing stories of playing in mud and her first kiss. The parents had long since left the young ones to party.

The music was loud and pumping the most recent radio favorites, and there was a small group of men in one part of the dance floor that started having an argument. Apparently one of the men had made a flirtatious remark to another woman who was at the party, but one of his friends didn't take too kindly as the friend had also been interested in the same woman. The two men starting hitting each other and one picked up a glass and smashed it on the other's forehead. It happened about three feet on the other side of the bar and within 43 seconds he was on the floor bleeding from his head. I called security and threw all 98 pounds of myself over the bar. We got the bloodied man off the floor and in to the back room and called an ambulance. I started applying pressure and wiping the blood off his face when I asked what had happened. He tried to brush it off and said the man that had cut his face open was his friend. But no friend of mine would have ever done that to me. I got him to medical attention and the rest of the party hadn't even noticed the incident.

I don't like to watch people tearing each other apart while under the influence. I've had insults of every variety thrown at me from every angle. I have seen good friends get lost in substance and circumstance and become complete monsters, and I watched that happen to my husband. Drugs, alcohol, strippers, hundred dollar t-shirts. It seemed super- glamorous for a while, life filled with super spendy trendy hair styles done by super spendy trendy Frenchmen beneath cut crystal chandeliers, tipping with bottles of French champagne. Beautiful outfits, beautiful shoes, beautiful things. But a whole lot of stress, blame and shouting matches. Nights spent alone in bed crying, days spent alone sorting out every facet of a thriving business; taking care of payroll, ordering, restocking, accounts, city council issues and a tax system I had yet to understand. Alone, just completely alone, and usually feeling overloaded with anger and resentment.

This feeling was only exacerbated one evening when I had been tending to customers and I noticed an unkempt man walking through the well-dressed crowd and hastily out the door. I hadn't noticed him inside the venue, nor had I served him at the bar, but I did see him leave. I walked to the back area where we had our office and noticed my purse had been opened beside the desk. My wallet was missing as was everything inside of it. All of the bills I had collected and held on to from my time travelling were all gone. The Australian dollars, drachma, lira, francs, yen etc., all gone, as were my driver's license, social insurance card and every other piece of identity I owned. I was so angry and felt completely violated. I called back to Canada at 3 am local time to try and figure out how to get my documents re-sent, and I was told it was nearly impossible as I was out of the country. I threw the office cordless phone against the wall in complete frustration and broke down crying. I had been robbed and couldn't even replace anything. I also couldn't apply for local ID without my Canadian ID. I called banks and various companies to cancel all of my accounts and put holds on everything else, but it didn't matter as they had already managed to steal my identity and opened other accounts.

It took a couple of weeks but I eventually had the identity theft sorted out, but I had to wait months until I could reapply back in Canada for proper documentation. Back home it was easy to replace everything, just as everything back home seemed easier. It all seemed harder also because I felt like I had such little support in Melbourne. Everything I did for myself, I did alone. Anything that Greg wanted to do, I had to do with him. I had to be involved in every facet of his life, from arranging his activities to paying his parking tickets or doing his laundry but if I wanted help with anything he was not even the slightest bit interested.

That overwhelming loneliness spilled over into every part of my life. There was a moment as I walked through a mall on my way to the bar that I spotted a puppy in a pet shop and stopped by to hold it. I'm not sure what breed of dog it was but it was tan in colour, had little floppy years and the cutest tail. I stood in that store and hugged the puppy for what seemed like an hour. Tears flooded from my face as he licked my cheeks and filled my hands with his fluffy warmth. I found it almost impossible to put him down and leave. I was desperate for attention and affection, but the attention I received from my partner was not that which I needed nor desired and usually came in the form of more epic verbal battles full of hateful rhetoric and blame. To be fair, I was an equal participant in these battles. I yelled back. I felt disconnected to everyone including myself. The carefully constructed interwoven web of lies that man was able to construct is actually a feat I thought sure none other would be capable of. There was a story for every person in every situation and in any given moment.

It was enough that I felt lost and unwilling to speak much of the time for fear I would not be able to keep up with whatever plotline he had been hatching. This worked for him as it gave him some solace and control in an otherwise chaotic existence. But it was not for me. I stopped enjoying even a glass of wine after a long day, and I felt no satisfaction from a pricey meal or a beautiful new outfit. I concentrated on what I could control, which was mainly trying to keep the bar functioning while my husband "entertained clients." I felt as though no one really knew me anymore, least of all myself, nor would even understand the gilded cage I felt myself to be trapped in. Everyone told me how much I had "the life" and I tried so hard to believe them.

One night I went out with my girlfriend to a trendy bar that was built in an actual laneway, with couches and chairs scattered along back staircases and fire escapes. We had joined with a group of her friends and were chatting on a couch when I began feeling dizzy and lightheaded. I had only had a few sips of my drink when I knew something was not right. It was my first foray with vertigo from stress and I had to leave right away. I could barely walk out to the street to get a cab, and I am fairly certain she had to carry me most of the way. I was lucky to be with her as she took me back to her place, tucked me in to her bed and provided me with what felt like the only nurturing night in four years.

I am so thankful for that night with her even though I was beyond embarrassed for having to leave her friends because I was sick. I could tell she wasn't terribly impressed at my physical weakness, but she took care of me nonetheless. It wasn't even as though I was physically ill. I did not have the urge to vomit or anything like that. It was that I was so emotionally unwell my body was reacting with a complete and utter shutdown. I was always so full of anger and sorrow, I was constantly fighting my own belief system in order to maintain the precious balance that was my life at that time and my body just couldn't continue. My inability to recognize my own needs was making me physically incapable, which also led to frustration and more feelings of inadequacy. Around and around it went, with no end in sight.

I should have probably just walked away from working at the bar and gone out to pursue my own career, but I didn't know what my own career would look like. One of the reasons why I decided to jump head-first in to the business with my husband was because I didn't know what else to do and also to make sure we would see each other because I thought if I had a day job and he worked nights we would never see each other. Even though we worked at the same place we hardly saw each other anyway. Sometimes things will go as they are meant to go, despite how anyone wanted them to be. It wasn't my career and it wasn't my life, but I tried to make a go of it. I took courses in design and first aid to do what I could to quell my need to grow and learn things for my soul but I still couldn't sleep at night, for a whole swarm of reasons. My relationship was faltering, I was working 20+ hours a day, and I had obscenely bad allergies year round.

I was having to deal with stress from work-related issues coupled with constant unrest with my partner, the gossip of infidelity, the inability to have meaningful connections with locals, the drug and alcohol use of everyone around me. The lies I was telling myself at that time were that this was actually the life I wanted and so it couldn't be terrible, it must be me that was making everything so terrible in my own brain. Maybe if I just loved him a little harder and supported his toxic lifestyle everything would be fine. If I dressed prettier, acted more the way he wanted me to, and learned to play by his rules then truly, this life would be just so great. I couldn't stand how dishonest my life had become, I wanted to crawl out of my own skin, roll up in to the fetal position and forget the world.

I held on to a few things that I felt were fundamental to what represented who I was as a person, and acting was one of these things. I had always kept one foot in acting in one form or another throughout my entire life as I still to this day feel very comfortable in front of a large group of people. I had spent some time on the indie scene working on student films and plays. One of these films was a dramatic horror short film written and produced by a group of

RMIT graduates. I had auditioned for the role and had been the successful candidate.

On the first evening of shooting, I found myself laying in a bathtub on a constructed set in an elementary school. The tiles had all been smashed with a hammer along the walls of the bathtub and had been smeared with black and grey paint to give it a dull, worn look. In the scene I was supposed to wake up under water and yank myself out in bewilderment and horror, not knowing where I was or how I had gotten there. On the third take, I had gripped the right edge of the bath tub at the exact spot that the tile had been broken and it proceeded to sever off the entire tip of my pinkie finger on my right hand. I screamed in genuine pain and it must have made for the most monumental of shots.

Blood was pouring out of my finger as I climbed out of the tub, soaked and freezing, trying to hold on to what was left of my finger. The young men in production came rushing over wide-eyed and pale but they offered me no solace. I asked for one of them to try and find the tip of my finger so they could reattach it at the hospital, but they clearly were not up to the task. I was rushed to emergency where I was treated and picked up by Greg. They sent me home with pain killers and a gauze bandage that they had carefully wrapped around the tip of my digit like a dog cone. They said there was not much to be done due to the placement of damage.

I found out later that night that the analgesics they had prescribed made me violently ill and very dizzy, and they did not allow me to sleep and gave me nightmares so I stopped taking them. A few days later my entire right arm went stiff and seethed in pain, so much so that I had to again be rushed back to emergency in the middle of the night. There was a young rather handsome resident doctor attending to the ER that night and he winced in empathy when he saw my mangled finger. He spent a few hours sticking needles into my hand in various places to try and block my brain from feeling the pain from the injury as he ran intravenous antibiotics through a vein my left arm. I was screaming in pain as he injected me over and over again but I couldn't say if it was more painful to have the needle or from the actual injury itself. I had gone septic and my body was not able to fight off whatever germ was now invading my body from its portal of entry. He sent me home with oral antibiotics and suggested wearing an oven mitt over my hand to try and protect it from knocking in to the bed or walls while sleeping. It was raw nerve endings, bone and flesh. To this day it remains overly sensitive, so much so that I usually keep that nail longer than most to give it an extra degree of protection. This injury was probably the first time I really had to consider no longer being involved in this world of Greg's. Eventually I had to stop acting because I found I could no longer claim joy in playing a role in something that was seemingly a dishonest portrayal of who I was as a person.

At this point I fought daily to maintain my sense of self when surrounded by contrived stories and circumstances. I also was now considered an immigrant as I had received confirmation that I had been granted permanent residency status. I was still called a dirty backpacker by most, and it didn't seem to matter that I had a husband and a business and a life. My accent that had been muddied by British slang was unmistakably not Australian. I was always labelled "too friendly, too outgoing, and too flirty", all things I ascribe to as being uniquely Canadian and were all seen as negative traits by many around me. I don't know why it bothered me so profoundly to be taunted, perhaps it was that I believed their summation of myself to be in some way true.

I tried my best to assimilate and I bought nice things and smiled when told to. I learned how to socialize and be a socialite, which was not such a far-reaching concept to someone so intrinsically outgoing as I am, but there was a certain level of shallowness this environment fed on, and I fed into that because I thought I had to. But I can't do that anymore as my solace lies in my transparency and I find comfort in being honest with myself and with others. I feel it is of great value to be able to look at myself and accept myself, to know me for my faults and flaws and still think I'm doing okay.

It is hard to look at oneself in the mirror and genuinely appreciate the person looking back, and many of us struggle with this. I am careful not to get lost in myself or my greatness as I have a strong humbleness and I understand all of the work that needs to be done in order for complete and lasting inner peace which is a dynamic beast that is ever- changing. I have zero desire to go back to a life that is anything less self-aware than that of my current soul. I know it was me in that situation that was not happy. I know it was no one's fault I felt the way I did given the situation I was in and I take full responsibility for that. I chose that life, it just happened and it was not the right life for me at that time. I know people did their best to be themselves, and I was the lost one. I had not identified what I had truly wanted or needed out of life, and I suffered immensely due to this unawareness.

But people don't know what they don't know, and I didn't know how lost I was. I didn't know how to find myself and I didn't even know where to begin looking. I know many people are like this, probably more than would care to admit it. Some go through their entire lives not even knowing why they feel so empty, and to question it is daunting and seems insurmountable. I don't even think I understood the plight of all of my life lessons until much more recently. Life has literally left me with no other option than to face myself and my emptiness in order to become a whole individual.

My entire life I thought that I would finally feel wholeness once I had coupled with someone else, but it seems the opposite is true for me. I felt less like myself when coupled with those I have found myself with. Moving forward I don't ever want to be left to feel so alone in a relationship that I don't even enjoy my own company. I don't want to listen to someone else's lies and excuses even when they are lies they tell themselves in order to sleep at night. At this point it is almost harder for me to listen to the lies people tell themselves to convince them that they are who they want to be, more so than the lies they tell me in order for them to convince me they are who they think I want them to be. Most of the time I still disregard my acknowledgement of the lie because I don't want to believe it to be true that someone would feel the need to fabricate anything for my benefit, let alone their own.

My thought on why people do this so readily is a fear that they will be judged for their perceived inadequacies. For the longest time I felt extremely inadequate and found it hard to just simply be myself, so I would bend to conform to what my partner wanted me to be. I wanted to be liked and loved and accepted, but that came at an expense for which I paid dearly. I have also discovered a small percentage of people that lie ostentatiously because they actually truly believe in that fabricated sense of reality. They may believe, for example, that the sky they see is actually purple, and they will try to convince everyone else that the sky is purple regardless of who or how they hurt another by doing so. In their world a purple sky is success and they will stop at nothing to paint all the skies purple for their own benefit. However with much patience and a lot of time truth always eventually comes to light and we begin to realize the sky is actually blue and the truth in the lie presents itself.

Nothing saddens me more than the moment their actions tell me the truth within their lies. I have learned that not everyone can handle the actual truth that is the blue sky, and there is no point in trying to get them to see blue instead of purple. To them it will always be purple. I have chosen to seek those who are more prone to being comfortable in their own truth more openly than I once did. And for those who want to embrace the blue sky, I help them find their own path to seeing their own true horizon--but only when they are ready to do so. If they want to hold steadfast in their purple sky, I let it continue to be purple for them but not for me. I have stopped accepting other people's beliefs as my own, especially when they do not align with my own values or they are less than honest or truthful. This is one reflection of the peace I have chosen to attain for myself: I am comfortable with who I am and by changing myself, I am changing my world and I am living under my own sky that is bright blue.

I knew the marriage was most likely ending quite some time before I consciously made the decision to end it. I was walking through a grocery store with my gorgeous girlfriend Bella back in Canada during a holiday, quietly sobbing and munching on a not-yet-purchased opened box of Arrowroot cookies I had grabbed off a shelf while following her blindly down an aisle. It was the only thing I had been able to digest as the anxiety and dread that had settled in my bones after realizing I would inevitably have to return to Australia to deal with things. My life was terrible and things had to change. I had tried to ask Greg to "maybe drink less" which resulted in him getting unbelievably defensive, hostile and emotional at first. After he yelled and cried at me, he said he would try but deep down I knew it was not possible, and within a few weeks that was clearly evident.

When I flew back to Australia that summer from my hometown, I knew it was over. He wasn't at the airport to pick me up when my flight arrived, and when he did get there sometime later his body was sweet with booze, probably still drunk from the night before, and he told me some elaborate story about why he was hours late because he stopped at a florist to get me a red rose (the kind women wrap in cellophane and sell for $5 at our bar every night) which he handed me as he went to kiss my cheek in greeting. "Uh-huh, oh wow thanks" I responded half sarcastically, wondering how to handle this. My stomach churned and my shoulders became stiff on the walk to the parking lot outside of the terminal.

I asked him how his time without me had been, and then sat in silence listening to him fabricate details and events for my benefit during the drive back to the city. It was now night time and we went straight to the bar to continue on with life as usual. I walked in the back entrance once the car was parked. I saw more cellophane roses strewn about all over the office desk and stock room and wondered how many women he had bought these for in a less than sober state. I greeted the bartenders who wore the uniforms I had created and scanned around the crowd as I did every moment I spent in the bar. The DJ was already playing music and the night was slowly beginning to grow. I sat down beside one of our longstanding employees and a friend of mine. Her sister had died a few weeks prior in a horrific accident and she burst on to tears in my arms. I cried with her for everything she had lost, and everything I knew I would have to choose to lose. She knew I was unhappy, and I knew she was devastated. Greg had already left to go "entertain clients" at this point and I was happier he wasn't there despite not seeing each other for weeks prior.

It was one thing to identify a problem and certainly another to do anything about it and I was still completely consumed with fear leading up to breaking off my marriage. It took me six days of anguish to make the decision to leave once I had returned to Australia. Nothing had changed and I didn't know what to do or how to go about ending this relationship, the bar, my current life, my career. To stay in Australia I felt would be very difficult, as I had found every moment of my life there to be overly challenging. Once we had separated, how would I go about rebuilding my life in a country that was not my own? He knew everyone that I knew and he was a much more prolific persuader than I. I would have to carve out my own career, my own friends, my own place to live all over again and dodge whatever bullets he would hurl my way after I would have so publicly rejected our union. I now thought that this was the most painful, most soul destroying time of my life. A broken failed marriage, all that time spent building a life with someone I thought I knew. All that time and energy spent to making that business something to be proud of, gone in an instant. Everything would be gone because I had failed miserably.

We both cried in the moment that I let him know my plan to move back to Canada. The pain was palpable and stifling and I second-guessed everything when I saw how much it cut him. I could not stop his hurt or my own and it all unravelled very quickly. He wouldn't speak to me for days and said all kinds of very painful things. He said I had cheated on him and used him for residency, even though I still have not been back to Australia since the day I left. I completely understand the pain and the lashing out of having your partner leave you though. I was prepared for the vitriol but it didn't make it any easier to receive.

I sent two small boxes home by freight and packed one suitcase, leaving behind every other material possession and memory. I only have one photo of my four years in Australia, of the landscape at the Twelve Apostles. I struggled for years with the guilt of a failed marriage under my belt, and it went around and around in my head for years following. I wondered if I had made the right choice especially considering how much pain and sorrow it caused both of us. I truly believed I could and should have tried harder to make our lives better even if he had absolutely zero desire to put in the effort I thought would be required to make a more healthy relationship. For him, our lives were perfect because he had everything he wanted, regardless of what that meant I had to give up and sacrifice including my values and morals. I felt every one of my friends judged me and thought less of me because I was so young, with a failed marriage under my belt. Some were less than surprised and thought I was a fool to have married him to begin with. They judged me to my face and behind my back. Most of my friends weren't even married yet, let alone divorced, and they couldn't comprehend or begin to understand the emotional rollercoaster I had been on. Many also had no idea what I was leaving and how hard it was for me to leave even though it had been so unhealthy. I felt incredibly weak and shameful that I was not the great success I thought I would be by that age.

Many more of my peers have been through divorce now, ten years after I left my husband. I judged myself pretty hard after the collapse of my first marriage, probably harder than I needed to, and it took me a long time--basically the entire 10 years since then--to be able to forgive myself. I feel strengthened to know I was able to have walked away from something in order to make myself feel whole. He did not want counselling, couples or otherwise, as he found his own way to cope with the loss of our union and I had to learn to be alright with that.

It wasn't all bad although at first glance it may look that way. It was the second step to finding out who I was and where I was going. Parts of our relationship were lovely as he taught me how to kiss well and how to enjoy sensual moments slowly. He knew me well, better than most, and I somehow understood the reasoning behind all of his pain and how he masked his insecurities, but it wasn't enough. He did not want to be truly vulnerable or honest with me or anyone else. Maybe one day he will find his peace, instead of only pretending to know what that truth might look like, and maybe he won't. I am grateful for the love he tried to give me, and the life he tried to build for us. I am grateful for the hard lessons I learned, and for even knowing him at all.

I honestly still love him and care for his wellbeing and I hear from him every so often. He does not lack for potential, and he does not lack for his own brand of love. But he was and is not for me. I could not make him happy, regardless of anything I did or did not do, and now I have learned because of other relationships including this one that I actually can't make someone else happy. Happiness is up to the individual, just as it is up to me to make my own despite whatever background noise may be trying to derail my path.

I don't blame Greg in the slightest for anything that happened. I don't blame him for anything he did or said or meant or felt. I don't blame him for our relationship not working out. There were two of us in the relationship and I take full responsibility for every mistake I made and everyone I hurt. I failed miserably at being his partner. He was doing the best he could, given his skillset, with a beautiful, quirky, headstrong, intelligent but utterly lost woman. I was everything he thought he wanted, but I was not what he needed. He was everything I thought I wanted, and nothing of what I needed. I honestly wanted to be the woman he wanted me to be, but it ended up confusing me even more as to who I actually was.

Through years of reflection and self-work I am able to see the gift that was a very turbulent, and at times opulent period of my life. I have no anger or sadness surrounding the experience. I am grateful to have gotten to know this man and his gracious family and friends, their culture and the culture of the Australian people. I was able to experience a very vibrant and always growing business, and I was able to experience extreme urban living; including the drug use, gang culture and deception that came with it.

I do not miss having to kick countless hypodermic needles off the steps to the back door of the venue or finding young people overdosing in the alley ways. I do miss the friends he shared with me and the laughter and hugs he afforded me. He made the entire experience happen for me, and he helped me to become the woman I am today and for that I cannot find fault. We would probably have been better off as best friends, but maybe we just met at the wrong time. I hope I will know him for a long time, and am able to show him my gratitude for the times we shared together even if they were at times like rolling through cactus. Looking back, I know I could have handled things better but at the time I was only doing what I could do as well as trying to figure myself out. I was a complete mess when we met, I shouldn't be terribly surprised to see our relationship was a reflection of that. I hope he knows he was and is loved by me. Despite our ill-fated voyage into union, I am so lucky to have known him, even for a moment in time. And I am so grateful for having the strength to learn and to become who I am now because of it. He opened his life up to me and even though it eventually did not work out, I am completely thankful for the experience nonetheless. I know I had to endure the suffering and growth in order to evolve, find my voice, and find my life.

CHAPTER TWELVE
THE FALLOUT

My flight back to Canada from Melbourne had two stops along the way, one in Fiji and one in Hawaii. This was done on purpose as it gave me a few days to transition and sit in purgatory before I reached my old life back in my old town. I didn't even cry much in those days of island hopping, not because I was full of enjoyment and elation for being in the tropics, but because I was so emotionally exhausted. I spent my time renting cars, driving through villages, collecting broken and bleached corals along the coastlines. I drove through countless groups of families on their way to and from events in traditional clothing. Many kept waving me down, wanting me to join them in their celebrations only I did not feel like celebrating. I just wanted to sit in my emaciated skin looking out towards the South Pacific Ocean with the warm breeze on my face. I weighed maybe 43 kilograms and I looked like complete shite.

Food no longer tasted good and even the most beautiful sunsets over Hawaii stirred little emotion in me. I spent a few days in paradise, and I felt completely numb. One day I want to go back to these places and experience them with true emotion and true life and true love. I want to capture them in their essence again, with me in mine. When I returned to Canada I was an absolute mess for months. I had bouts of vertigo that left me crippled on the ground and unable to move for hours. One occasion my dad found me and poked at me as though he was holding a stick and I was roadkill. He checked my pulse and asked if I was still breathing. I was. Eventually the deep throes of imbalance subsided and I found a job, enrolled in college for the third time, and hoped my life would somehow get itself on track again. I had less than nothing when I returned, as I just walked away from everything and my family graciously invited me back in with open arms while I struggled to find my identity.

I spent some time dating the wrong people, and I dated some absolutely lovely people that were just at the wrong time in their lives, and me in mine. One evening while I was out celebrating a birthday with some female friends, a gorgeous curly haired man sat down beside me and introduced himself. He told me he was the son of a diamond miner in Yellowknife and happened to be in town on business. I asked what he liked to do most which turned out to be photography because as much as he thought he was charming me with his fanciful tales of jewels and luxury, I just wanted to know someone in their simplicity. He could not possibly know what I had recently been through.

After a few weeks of chatting he explained that he was living in Calgary and shockingly it came to light that the diamond mining story was not true, so he sent orchids to where I was working as a receptionist in a high-end car dealership to apologize for thinking that is what it would take to impress me. We long-distance dated for a short time and have remained friends through the years. He has a gorgeous child and spouse and has since survived cancer, thankfully, and spends his time creating brilliant products from different types of wood. Part of me is still impressed that my diamond miner thought he would have had to be the heir to a diamond mine to get my attention.

School wasn't going all that well. I had enrolled in a business program to have some academics to strengthen the work I had done overseas and to strengthen my resume. I found the classes less than inspirational and barely received a passing grade in some of them, especially accounting. Apparently I overcomplicated the process of debits and credits to the point that I didn't even understand myself or how something could possibly be so easy. I commonly do such a thing, thinking that something must be way more complicated than it actually is. I think the prof just passed me out of pity, although I know they don't truly do that kind of thing.

During one marketing class the instructor started explaining as per our textbook just what social responsibility meant in the business world and in the broader sense. He described how a business is obligated to keep shareholders happy, with profit margins being the most pertinent thing we had to do. I totally understand this on a basic capitalistic level but I just thought there must be more to business than driving profit up with no real purpose other than keeping investors happy, so I started to doubt that business and marketing was the occupation for me. I do see the need for maintaining fiscal responsibility as a business but at the time I didn't feel the heart or soul in my current scholastic endeavors.

I stopped trying and stopped putting in effort because I knew I would not fill my cup with this path. After meeting with friends, colleagues and guidance counsellors I decided to enrol in nursing and try a completely different path of education with a different, more humanitarian purpose. After all, I seemed destined to care for others without much regard for myself so it seemed to make a lot of sense. As a child I thought I would have liked to be a psychologist or a doctor, but never actually thought I was good enough to get through the schooling to attain these seemingly lofty goals. I had waited months to hear whether or not I was to be accepted into the program which had a large wait list, but once I heard I had been accepted I knew the path I would now be on and my short term future was certain.

The program didn't begin until the following fall so I decided to take a few weeks off and travel around Mexico with a friend who had been living in a small town for some years. I quit my largely dead-end job and cashed in some of my savings for plane tickets and bikinis. My friend had sent a list a mile long of supplies to bring down for her which I dutifully filled with the help of her family and one very large pink suitcase of mine.

The flight to Mexico was largely unremarkable. There were two connections that went along smoothly and I found myself quickly exiting a plane on a rural tarmac in a very humid, very lush tropical land. Airport security was a small room basically outdoors and the attendants had open-carry rifles that hung across their chests. It did not make me feel welcome. This was my first foray into Mexico, and it was quite the sight to see guns on everyone. They went through all of my baggage, not seeming to mind that I had a few thousand dollars' worth of hand tools and beauty products mixed in with housewares and new clothes. Most of the other passengers had brought surfboards in large surf bags and had shell necklaces around their necks, with an anxious look in their eyes as they patiently waited their turn. I was in the land of the surf bum, and I couldn't be happier.

My girlfriend was waiting for me inside the terminal and gave me a huge hug upon seeing me released from customs. We quickly caught up and loaded my luggage into her white Astro van and cruised off towards town. This town was known as a surfer's hidden mecca until recently. It runs along a bay of pristine coastline that many locals and foreign nationals flock to in search of the perfect wave and a paradise to call home. She took me to her favorite shops and restaurants to acclimatize me with the geography and layout of the town. I ate fresh street meat tacos complete with the obligatory hundred hovering flies and squeezes of fresh lime. I ate mangoes directly off a mango tree in their yard and cracked open passion fruit to slurp down for snacks.

I stayed in a small one-room thatched roof cabana off the main house that had its own walk in tiled shower and sink, with a small food preparation area and hammock outside on the porch. It was surrounded by lush greenery and a small man-made pond at the front full of frogs and other domestic creatures. The property had a gate along the roadway and numerous fruit trees around the perimeter. I helped my girlfriend to hand-wash clothing and linens and hung them to dry in the warm afternoon air. We spent many days on various beaches, drinking local beers with friends of friends and having conversations about nothing in particular.

I had a surfboard custom made in honor of my trip, complete with a Hawaiian phrase written across the front that I chose from an old manuscript I had found that meant "a gathering to signify peace." I have always loved Polynesian culture, and I thought it suiting to have such a beautiful phrase painted on my beautiful pink wave rider. I have always pursued peace, although until now it has eluded me. I have learned in time and through much pain that external peace is only achieved after internal peace. Although I have read this enough times to know it is true, it takes struggle and dedication to learn how to embody that on a daily basis and very few continue to be successful in embracing it.

I took my new surfboard to a few of the quieter beaches with smaller waves and clumsily attempted to paddle my arms fast enough to get out past the break. This took some time as I was not in the best upper body shape. Eventually after inhaling a few mouthfuls of water and getting thrown backwards a few times, I managed to push past the break and into the stillness of the ocean. There were half a dozen or so people out on a hot and sunny day in June. I managed to catch the fourth wave that came by, but so did a young local man riding a very short surfboard who seemed absolutely disgruntled at the thought of sharing this moment with me, so I threw myself out of his way to avoid a nasty confrontation.

I paddled back out and sat in waiting for a wave that a local wasn't already getting. Something I soon learned about locals and "their" waves was not to drop in on them, ever, even if by mistake. It seemed pretty silly considering no one owned the ocean and I wasn't at the beach that was considered for the quasi-professionals, but I obliged regardless. I patiently waited and watched the deep water under my feet and at last I did manage to catch a wave all on my own and rode it all the way in to the shore on my feet without breaking any bones. It was a triumphant occasion, one I shared with my female friend as we were packing it in for the day. I have surfed occasionally throughout my life but have never had the drive to become any kind of dedicated athlete. I do it because I enjoy being out on the water.

There were thatched-roof cabanas along the shore that sold drinks and snacks and we had just cracked open a coconut water and some fresh guacamole for our chips when another tourist came up to us, hobbling through the sand with a surfboard under his arm. I looked down and noticed his right foot was absolutely covered in blood. He had been thrown from a wave up onto the rocks along the northern bank of the cove, and when he gathered himself to climb down to the beach he had stepped squarely onto a sea urchin whose spikes had shattered within his foot. He was a pale-skinned, tall British-accented man who was trying his best to seem unmoved by the excruciating pain he was in. I suggested he visit the hospital as they would have to try and remove the broken bits that were now deeply embedded in his flesh. It looked like a horror scene, and we offered to help him carry his things up the high vertical steps back to the parking lot above the beach, but he embarrassingly declined. We winced and turned back to our snack after watching him precariously make his way upwards.

Once I felt I had become stronger at my surfing I decided to go to another beach which was walking distance from my girlfriend's home. I packed a towel and a bottle of water and threw on a bikini and rash guard and headed down. This beach has some notorious waves with some fairly impressive power behind them. They have a tall and swift break, and for some reason I felt like I was ready to at least get by. I chose the opportune moment to paddle out past the break and made it out after only a few attempts. The swell was pretty big by my standards that afternoon, well overhead and my board was big and awkward to press the tip below the crest of the wave in time to not get stuck in the spin of the wave. It wasn't until I made it out past this point that I realised just how large these waves were, and I found myself wondering how I would ever make it back to shore without being swallowed whole by their power. The wind was offshore and there were dozens of other very skilled surfers now gathered by the rocks in search of a perfect barrel. I sat straddling my surfboard, my chest pounding trying to figure out how to get back to shore. All of these surfers were the kind who did tricks and flips and spins in the air above the surface of the water and easily jumped from rocks directly in front of towering curls with ease. Some were as young as seven, shredding the snot out of the face of larger waves than I had ever seen this close.

I saw a lull in the surf and decided to try and paddle back to land before the next set even began to approach, but it was too late. I was only midway back to shore when I could feel the pull of the ocean on the air against my face, and the back of my board began to tip up and back as a wave grew behind me. I paddled harder to try and keep up with the force but there was no way I could have. The wave ripped my surfboard from under my belly and sent me head first under the water, deep down almost to the ocean floor. As soon as my surfboard caught the current of the wave it sent me straight upwards like a ragdoll back towards the surface. Instead of breaking the surface tension, the drag of the board followed the wave back down towards the ocean floor. I could feel my heart racing as my brain realized I was quickly running out of oxygen. My body was star-fished and being tossed wildly in every direction almost simultaneously, the leash of my board now tightly wrapped around the length of my thigh cutting off the circulation to my foot. I clumsily fumbled around to reach the ankle strap that bound me to my surfboard and tugged hard. Instantly I was released from the grips of the machine washer that was the wave.

I scrambled to the surface and gasped for air. As I went to take a second breath the second wave hit me from above, sending me deep down to the ocean floor again, but this time I was able to grab a handful of sand and crawl across a few feet of the rippled ocean floor. I looked up and saw the wave delightfully curling in the sunshine above me. For a brief moment I thought I would drown as my lungs became tight with their yearning for gas exchange. I dug my feet in hard to the sand and propelled my body towards the surface in a desperate attempt to feel the warm air on my lungs again as the muscles of my body throbbed in oxygen deprivation. I gasped for another breath of air and quickly jilted my head around to see if I could find my surfboard, in hopes that it wasn't about to come crashing down on my head, driving me unconscious and complicating my situation further--but it had already made it to rest upon the shore.

I flailed and kicked towards land and just as I felt the shore sand in my hands, another wave came and dragged me back down, pulling me inside of it. I clawed at the sand to slow the momentum of the drag and eventually got out of the never-ending cycle. The last wave threw me back on shore where I crawled for a few metres, gasping for air, in the direction of my surfboard. An older man was sitting up along the sand and had watched the whole thing. When I approached him with my surfboard under my arm and a large swirly bruise wrapping all the way up the entirety of my left leg he calmly asked "what happened?" I muttered something about raw power and sat down in the hot sand, gathering my thoughts and my breath and feeling so embarrassed that I had even attempted to ride the "big boy waves" of La Punta. That was a close one, I silently thought to myself, and another terrible decision. It was not the first time the ocean tried to claim my life, but I am hopeful it will be the last. I spent the rest of that day and into the night sitting in much stillness, contemplating the scene I had survived. I visited with my girlfriends friends in their open walled, open concept homes and listened to them tell stories of past adventures. My body was tired from my day and my leg throbbed from the leash bruise so I retired early.

A couple of days later my girlfriend and I had planned to go out for a night on the town, however I ended up just going out with the brother of my friend's boyfriend to take in a night of discotecas as she was caught up in a home life issue. We stopped at various bars and clubs and I tried my best to keep up with the quick footed locals in Salsa and Merengue. We each had a few drinks under our belt as we made our way through town to other venues. He spoke some

English, and I spoke some Spanish but the conversation was fairly sporadic and basic at best. He introduced me to many of his friends and even went so far as to introduce me as his novia (girlfriend) which I thought was a rather liberal step in categorizing me as such seeing as we were hardly friends, let alone lovers.

I had worn my hair down for the majority of the time I was in Mexico as the humidity coupled with coconut oil had managed to form rather perfect curls when usually my mane remained temperamental and unkempt, and that night was no different. I had on a tight grey denim mini skirt and a pink halter top with an open back. He enjoyed coyly sliding his hand around the side of my abdomen as he introduced me. I was not interested in this man in the slightest and I thought I had made that rather clear to him privately so as not to offend his ego, but it didn't seem to matter. By the time we hailed a taxi for our journey home we were pretty tipsy and sticky from humidity and he had started taking liberties with my face in his, kissing my cheek and chin passionately. He was trying his best to whisper drunken Spanish sweet nothings in to my ear, but I was really doing my best to ignore the fact his mouth was making out with my chin bone.

I carried on a conversation with our taxi driver the entire way home as he continued his assault on my face, this time with his tongue firmly travelling to the depths of my inner ear. The taxi driver found it mildly amusing at my nonchalant approach to dealing with this rabid human. I was careful not to offend, however, to ensure that my personal safety would stay intact. I thanked the cab driver and my companion for the night as he asked repeatedly if he could join me for an overnight stay. I thoughtfully declined about 17 times before simply walking away and shutting the gate behind me. The next day he had gone ahead and taken liberties to describe in great detail the amazing, apparently tremendously romantic evening we had together to his friends and anyone who would listen. It didn't matter that I hadn't even kissed him at any time.

I had noticed a few weeks into my stay that there had been a young man next door who had developed a bit of a cough. He spent the majority of his days sleeping underneath a tree in the backyard, set far back from the main house that housed a family with   numerous children. He looked gaunt and pale when I began to take notice of him. His cough sounded deep and hoarse. I asked my girlfriend if he was dying of cancer and she didn't think much of it. I asked how long he had been like that and she couldn't say. I suggested to not let her child play with the children from the neighboring house as it sounded to me like it was either cancer or tuberculosis due to the heavy nature of the sound, and the fact the man looked emaciated and frail even though he couldn't have been more than 21.

My girlfriend tried to explain to me how education does not exactly take precedence and many of the locals would never dream of going to see a doctor, but I urged her to ask them to get him some medical attention. While I was out meandering town the following day she had printed off some information from the internet and had read to the mother about tuberculosis and how he might die if he was not treated properly and in time. She could not read, nor could her son who was coughing continuously under a tall palm tree. He did go off to a doctor and was indeed diagnosed with tuberculosis. I am just happy it was remedied and not passed to the other children in the village.

During my last week in town we took trips to a turtle sanctuary to see hundreds of turtles at various stages of growth, including some very large older turtles. We also went to a few popular beaches south along the coast and devoured numerous dishes of fresh seafood. I did happen to have food poisoning on more than one occasion although I would be remiss to identify whether it was due to the food, the water or the ice or anything else. Most of the discomforting symptoms of vomiting, stomach cramping and diarrhea passed in time--other than when I opened my mouth in the shower by accident and took a parasite home with me. Who doesn't like to take home an authentic souvenir?

We went to a very well-known restaurant to celebrate the birthday of some local friends on one of the last evenings I was in town, and stayed up rather late at a party hosted in a private home. My girlfriend left early, and I followed not much later. I had a few kittens staying with me in my cabana that were acquired as strays a few days prior, so I had left the door open just wide enough so that they could get out in the middle of the night if they so desired. A few hours into my sleep my friends' boyfriend appeared inside my door, wanting to "chat". He came in and sat very close beside me as I propped myself up against the wall behind the bed and he started saying and doing rather inappropriate things, so I gently told him it was time for me to get back to bed and to have a nice night.

He was rather hesitant and continued to make some moderately colourful physical advances to the point where I began to feel extremely uncomfortable so I again asserted that it was time for him to go. I put a kitten in his hand and waved towards the door. After some stalling, he did leave but it left me feeling very uneasy.

The next morning I was up early and laid in a hammock on the patio writing in a journal. I had mentioned to my friend that he had made advances in my direction that were not welcome and had let her know that I ushered him away post haste. I did not want her to feel that I would ever overstep any boundary of her relationship, nor did I want her to hear from some other person later that I hadn't told her.

A few hours later, a certain boyfriend slinked out into the daylight soaked courtyard and tried to apologize for his unwelcome behavior before I had even said anything. I scrunched up my face and gave him a disapproving look. I had no idea in that moment how much of a catalyst this turned out to have been. Things got pretty intense after this episode, as it turns out these weren't the first advances he had made towards a friend but possibly the first time they had been denied. The couple broke up rather spectacularly and I felt one hundred percent responsible even if the actions were his and not mine.

I took refuge with a neighboring friend who had already gone through disagreements between them and she assured me it would all blow over in time. Only it didn't, it escalated and when it was time for me to leave I felt very afraid for her. I tried to give her all of the pesos I had not already spent so that she could go to a neighboring city and stay at a resort for some reprieve, but she tearfully declined. Upon my return to Canada things only got worse, and she eventually had to leave. I am thankful for her safety, but I still feel terrible for being the eventual trigger that ended a relationship, especially considering it was because I did not want to sleep with someone, not because I actually did. Our mutual friends took sides in the debate, and I was not the popular vote. But I am okay with that because she was not safe and she was not being treated with respect. I will choose the unpopular vote if it means standing up against things that aren't right. I still have so much love for her.

On my flight path home I had quite a few hours to spend in the airport in Mexico City. I started chatting with another man who had a similar wait to Los Angeles and shared with me that he had recently separated from his wife and was in the throes of divorce. He had decided to take some time off and go on a surfing trip in Mexico which he was just returning from. He was a teacher from a smaller city in California and it sounded as though the separation came with much angst and regret. He had three sons all of whom were avid little surfers, much like his dad. They had just finished renovating their family home, and his wife had undergone cosmetic surgery before it all fell apart. He remarked at how calm and collected I had seemed, as though I was full of a delicate lightness when we sat and spoke. In that moment I did feel that a stillness had fallen over me. I was just returning from almost two months of holidaying on my own and was looking forward to the next adventure that awaited me upon my return home. I let him know I was sorry for the breakdown of his relationship, and how tough that must be for everyone involved. We spent the majority of our layover conversing about whatever topics floated to mind. I had noticed after taking a trip to the washroom that I thought his gate and flight might have been changed. He went back to check the screens and had to dash off in a mad rush to catch the flight that was now just minutes from leaving without him. We remain pen pals and he even sent up to Canada a parcel of his favorite tortilla chips for me to try. We are still friends and he and his sons have become very talented surfboard shapers and creators. It has been a remarkable thing to have such an off the cuff, unexpected exchange in an airport turn in to a lasting friendship that spans hundreds of miles. Just as with most of the friendships I have been blessed enough in creating and keeping all over the world. That I have not seen this person in ten years does not diminish the bond I have with him, or anyone else for that matter. I feel deep gratitude for being able to have and keep these connections regardless of personal circumstance.

   I continued on my flight home, mostly worrying about the future of people that were no longer in my presence. I reflected back on my time spent on holiday and began to try and wrap my head around going back in to full time school once again. I was not apprehensive at all for what lay ahead of me, as something deep down inside of me knew this was the right path and the right step for me, finally.

CHAPTER THIRTEEN
THE END OF THE BEGINNING

I spent much of the rest of the summer working and trying to prepare myself for a new experience, not fully knowing what to expect or even if I would find success. In the fall I armed myself with student loans and a backpack full of overpriced textbooks from the college bookstore. My nursing program had started and I had not prepared myself for the sudden shift in focus very well. I had been seeing someone loosely, but he was really not all that great for me so I found studies to be a great and valid distraction to having to deal with the rejection and moving on from an undesired coupling. He is and was a fine human being, but incompatible with me in the long run and I had at least come to identify this at long last. I felt like I was always his "fall back, in case of" woman, and I deserved better than that.

The first week of any new course or program at a school institution can be overwhelming at the best of times. I had to read multiple chapters in numerous books on a nightly basis but none of the content was terribly hard to grasp. I had learned what my best study habits were from my youth and came to perfect them during this course. My classmates were varied and some I had even known from random past events and high school. I was still the bubbly, sassy one at the back of the room, and I had to try hard not to make overtly sarcastic quips. I started dating again, but for all the wrong reasons, and it only led me to trouble. The men I dated were not the best of characters, nor did they have the best of intentions and I decided somewhere in the thick of winter to completely abstain in order to focus on grades and learning.

Studying nursing and health seemed to come naturally to me. Once I learned the basics about cellular structure and homeostasis everything else seemed to makes sense. Homeostasis is kind of a big deal, as anyone that had me as an instructor in my later career would know. On every level and in every sense our bodies are constantly trying to be in balance. If you put this same concept into all aspects of life, everything makes sense and nothing seems as complicated as it once did. Everything is trying to constantly get back in to balance with everything else. Just as our bodies strive for optimal health always, so does everything in nature. If it is too hot, the body tries to cool itself down by way of sweat and if it is too cold we have goosebumps to warm us up. This concept is as far reaching as physics, as for every particle there is there is also the compliment particle to that particle.

When I reached the halfway point in my schooling I also received the most unexpected letter in the mail from Australia that showed my divorce had been granted. I have no doubt this was mostly due to his new girlfriend taking the lead in filing the application as he was known not as a doer to completion kind of a guy but more as a talker. I had been told previously that international divorce was a tricky chestnut and it took years for me to figure out the process. I learned that legally a divorce wasn't really necessary if I had no intention of pursuing support or a payout of any kind, except that it would hamper my ability to be legally remarried. I was relieved to receive that long form piece of paper after much fretting about it. It allowed me to let go and feel as though that chapter of my life had been completed. I continued on to finish the nursing program the following summer and finished with rather strong grades, graduating with much ease. Now that I felt I was on track, nothing seemed as hard as it once did.

I went straight to work on a casual basis for a few months before applying for a job in another area of Canada. Within 24 hours of sending my resume I had a job interview lined up and the prospect of full time employment ahead of me. It wasn't the best region in Canada for someone who prefers tropical climates and warm evening breezes but I saw the potential in a permanent full time position that would offer me loads of experience and knowledge. When I arrived for the interview it was in the throes of a very cold blizzard, the snow was multi-feet deep along the streets and the wind cut through my west coast coat with ease. The interview went well and management seemed keen to have me despite being fairly fresh out of school and I, with my trademark exuberance and enthusiasm was grateful to be given the opportunity.

At the time it was my most desired position, working in a surgical stepdown unit of a general hospital. I had decided prior to graduation that if I could get at least a year of acute care under my belt then I could parlay that experience into something better and more accommodating in furthering my career. I learned early that shift work was much too hard on my body, and working overnights turned me into a zombie. It took me two days to shake what a block of overnight shifts did to my body and my brain. I met some very good friends and also learned quickly those strong personalities it was best for me to avoid. I used avoidance early on in my career when I really did not know what I was doing and did not yet have the confidence to deal with bullying as well as my actual job, but I was bullied quite a bit. Nurses are notorious for "eating their young" as it were. There were those that were kind and helpful but there were always those that were unsatisfied with everything that found every opportunity to reduce a co-worker to tears.

One group of coworkers liked to shun or to speak derogatorily to those who made them feel threatened, and for the longest time I could not figure out how I was possibly seen as a threat. One particularly hard night shift I had a patient take a turn for the worst and we needed to transfer him to a more critical unit. I was able to participate in his treatment in the extent that I knew how, but when I returned to my unit a comment was made in passing about my body and suddenly it made sense. I was seen as having less value as a co-worker because I was thin and it made my coworkers uncomfortable. I know for a lot of people the perception is that only curvier women have troubles with being treated differently, but this has happened my entire life. I have always been slender and I have almost always been referred to as anorexic and unhealthy. Even as far back as high school I would show up at my doctor's office in tears wondering how I could gain enough weight to have the curvy body that seemed to be more widely accepted. I was so tired of getting bullied for having a small chest, or hips like a boy, and I just wanted to feel accepted.

Once I realised that this was exactly what was happening in my current adult professional life it made sense to me and I started making jokes to break down the walls of insecurity that my body had made others feel. I find it so frustrating that we continue to perpetuate stereotypes about what a body should look like or feel like or be like. I have never put someone else down because of how their body looked. Many people think I don't have these same issues that others in the opposite end of the spectrum face, but I do. It hurts equally as bad to be called anorexic as I imagine it hurts to be called overweight, and it never is a good thing to be compared to someone in a concentration camp, or to be told I could never possibly be found attractive because of the size of my bra, or because of the prominence of my collar bone.

When people don't love or value themselves, they try to bring down everyone else around them regardless of height, weight, colour or creed. I just decided to try to stop listening to others and instead start listening to my own internal rhetoric. I have many values that are not equated with the size of my clothes and I do not perceive myself to be of lesser or greater value than another human. I do make jokes about myself from time to time, but I now know I am a beautiful human being inside and out. It did make working with these few people a little less hard with the knowledge and understanding that it was their insecurities that were the primary motivation to their exclusion, but it didn't excuse their behavior. They all eventually moved on to other floors and I ended up getting pregnant by my then boyfriend.

My pregnancy was unequivocally very harsh as I was part of a very small percentage of people that had runaway nausea and vomiting for the entirety of my pregnancy. I had to stop working because I was throwing up in the bathroom on shift for the majority of the first few months. I found it hard to stop working, and I found it even harder to keep any kind of food down. Merely looking at the kitchen, reading a book or watching television would make me nauseated. Luckily I had bought a puppy a few months before growing baby and she made for excellent company. She was a gorgeous Boston terrier with ears larger than her head and the most loving ball of energy I had ever met. She actually chose me when I went to the breeder's house because when none of the other dogs paid me any mind, she would not leave my side. She was black and white and weighed all of 12 pounds as a full grown animal and was the brightest animal I have ever known.

My boyfriend was working out of town the majority of my pregnancy and so it was me and my dog largely on our own for nine months. She kept me busy with fetch and walks around the neighborhood and licked my feet as I threw up in the toilet. Because I was so sick and tasting every food twice like a mother bird regurgitating, I didn't ever really get to feel that pregnancy glow I have heard so much about. I was tired and had no appetite although I managed to eat and keep my baby fetus nourished regardless.

When my son came into my life it was in the dark of winter and it was a very intense labour. Everything that could go wrong did go wrong, despite my screaming and yelling to advocate for myself and my unborn child. He was whisked away the moment he popped out and was sent to the NICU to be monitored. It was a surreal experience, and not something anyone could take lightly. In speaking to many other mothers, there does not seem to be many that look back to child delivery with bliss and wonderment though. I am unsure as to whether or not I can say with confidence my son's harrowing transition to breathing air was any better or worse than any other birthing story. I am, however, excruciatingly grateful he is and was a very healthy boy.

The first year of my son's life was much the same as any other new mother's story I have heard in terms of adjusting to life with baby, full of mind-numbing tiredness while trying to juggle the surreal level of expectations that are thrust on new moms and I did my best. He ate organic food and dirt probably more than once but he was healthy, happy and well-loved. He probably even ate organic dirt, which is a step above normal earth dirty dirt I suppose. I stopped reading books on how to be the perfect mom early on, mostly because they gave me more anxiety about all the things I should be doing rather than whatever else their intended reasoning was. I was often raw with the stench of motherhood, that savory mix of body odour and baby vomit. I did get to shower occasionally and even took myself to a salon every once in a while. I give myself some credit for being a smart mom in that he didn't play with knives or plastic bags or weapons of any kind which is really quite important. I spoke and still speak more than one language in his presence, and he doesn't eat junk. I may not be the picture-perfect example of whatever is the most trendy mom picture there is at the moment, but we did alright.

About a year after the blessed calving of my son I was standing in my living room after I had fed and changed my child, sending text messages to my friend and co-worker who now lived in the basement suite of our house. She was giving me a heads up that she would be away for the weekend on a fishing trip to the family cabin and that I was not to expect anyone to be in her suite in her stead. I wished her a great time and continued on with cleaning up from lunch for myself and my child. A few hours later I received another message from her stating she was coming home and the fishing trip was cancelled. I told her I thought it was a shame and asked what had caused the change of plan. She messaged me back and I stood in shock as I read the message out loud to my then boyfriend. Her parents had died in a mid-air plane crash. I immediately burst into tears and was in complete shock as that was the furthest from what I had expected to hear back. I thought maybe someone had to go to a meeting or a dental appointment, or maybe the car had a flat tire. Not this.

She arrived home and was in complete shock. I just held her for as long as she allowed me to and made her cherry and grape jelly when she finally had the stomach for it three days later. Media had swarmed her family home so she took refuge in our backyard instead of having to deal awkwardly with questions she did not have answers for. Reading the story in the news was like an episode of some alternate reality as it does not happen too often that mid-air tragedies take place. If either plane had left 10 minutes earlier or later, or either had been flying 500 feet higher or lower it would not have happened. Loss is a bizarre thing and it creates all kinds of odd reactions. She was an amazing soul through the entire experience, and I honestly don't think I could have handled it with as much grace as she did. I felt like I couldn't do enough to give her love at that time, but she survived. She has since put herself through more school and continues to amaze me with her strength and fortitude years later.

The relationship with the father of my child eventually did not work out and we all decided to move back west where the winters were not quite so harsh. I found a new job, started a small business, and settled into a new home and a new routine. Nothing could eclipse the pain of the end of this relationship and its aftermath, but it finally led to my real awakening. I dated on and off for a while with one working professional who was very good at pretending to be happy and successful but over time I learned the cold, hard fact of his truth--his complete and total unhappiness and quiet discontent. He had a past as we all do, only he was not able or willing to shake it to live in the present. I actually thought he was very amazing, but it is a pity he didn't see that in himself. Much could be said about the father of my child as well, but as I have learned, I can only be responsible for my own happiness and my own actions.

Right when I was thrust back in to singlehood, something rather unexpected happened out of the blue at just the right moment. The Irishman from Australia sent me a message online. He was sitting at a pub in London, swilling a few beers with some mates, and thought it was the opportune moment to catch up and he could not have been more right. I may have let it slip that I had a large crush on him all those years ago, with his tight black V-neck t-shirts and well-placed leather wrist bands. He basically had the nicest bum of anyone I've probably ever known and oddly he apparently thought I had a rather nice behind at the time as well, so we carried on conversing for a fair few months. It was absolutely delightful to touch base again with someone I had known in another chapter of my life. He was smart, gorgeous, successful, very charismatic and very easy to hold all kinds of conversations with. He was now living in London and absurdly busy travelling the world being generally amazing.

One night around Thanksgiving I had gone out with my girlfriend Janice to a couple of nice restaurants and lounges in town and we had been chatting with a few groups of men. There were two young men from a small northern town to our left and a larger group of four or five men to our right that had come into the lounge. I had not had much alcohol in the eight or so hours we had been out and was sipping water at the time. The larger group of men were in to celebrate their married friend's 40th birthday without his wife. We went back and forth speaking to the different groups of men when I started to feel dizzy. I was standing up chatting with the married man whose birthday it was when he started to put his hands up my shirt and down the front of my pants.

This caught me off guard and a few moments later I began to feel nauseated. I ran to the bathroom as I thought I would be instantly sick. My legs began to get shaky and I thought I was also going to have diarrhea. I spent some time staring in to the mirror in the bathroom trying to count the drinks I had consumed at that point and decided it was best I go home immediately. I went back to my girlfriend and explained that I was not feeling well and had to go. She was less than sober and decided she would stay to finish her drink as my house was only a 15 minute walk away. I left the venue and headed for home. I was having trouble seeing straight, and couldn't even make out the crosswalk with much definition. I had only made it kitty corner to where I had left the venue when I could feel my heart begin to pound. I couldn't breathe and my legs began to wobble. I started to panic and scrambled to reach into my purse to find my cell phone. My legs gave out from under me and I hit the pavement face first. I crawled to a bench and struggled to dial for help. First I called the man friend I had recently split from, in hopes that he was awake at 1:30 in the morning but with little success. I then tried to dial 9-1-1 but it took me forever to mash the numbers with my slowly moving limbs.

The dispatcher couldn't even make out the words I so clearly heard in my head and it hit me that I had been drugged yet again, and panic set in. I was downtown in the middle of the night, alone and unable to defend myself. I put my head in my hands between my legs and began to pray that the ambulance would get to me before the guys who drugged me did. I propped my head between my legs and continually tried to raise my upper torso to at least make myself look like I was not a complete invalid. I did not want to look like more of a target than I already felt but it was impossible. My body was numb. Fortunately none of the passers-by took much notice of me and the ambulance arrived. By that point my speech was coming back and I could use my legs again, if only slowly.

They loaded me in to the emergency vehicle and the police arrived. The female officer pleaded with me to go to the ER to have my blood tested and to press charges, but I had no idea who I would point the finger at. The paramedic was male and unsympathetic, stating that by the time they would get around to drawing a sample it would be largely out of my system. They then let me walk home by myself. It was after 3 am and they should not have done that. I felt embarrassed and just wanted to get home, so I just started walking. As soon as I stepped off the ambulance, my Irishman messaged me and I told him what had just happened. He chatted with me the whole walk home to make sure I was at least not completely alone. When I arrived back at my house, Janice was in the yard wondering where I had gotten to. She had no idea I was flat out on the pavement, nor that I had been drugged. I said goodnight to the Irishman and thanked him for virtually accompanying me to safety.

My girlfriend and I tucked ourselves into bed and I locked the door tightly behind us. I was really rattled by this episode of public drinking being drugged. It left me feeling violated much more than the past two times for some reason and I am just thankful it didn't end up being a much worse outcome. I still don't understand why a man would want to drug and have sex with an unconscious woman. I don't understand the allure of invading someone in that manner or having complete and blatant disregard for consent. Isn't a physical relation supposed to be a bonding experience for two consenting adults? Why has it turned in to something someone can just take without any regard to the other person? The Irishman and I continued on for a few months having regular chats but eventually the distance got to both of us and we parted ways, but dang if I were to ever be reunited with that man even for a moment, there would definitely be a solid consensual hug or two to go around--especially if he wore a tight black V-neck t-shirt.

I did manage to date a few very interesting characters post-Irishman. One in particular made for probably the most incredibly grandiose first date of my life. This gentleman had introduced himself out of the blue to me on an online dating website. He was from another town not far away and did rather well for himself with a whole umbrella of companies under his belt. He had recently been divorced and seemed very nice and respectful. A few weeks of casual chatting went by and he invited me to enjoy an afternoon together. He picked me up from outside of my wartime-era home in his very fancy car and handed me a dozen roses, wrapped delicately and tied with multicolored string.

We drove to his house which was roughly 45 minutes away and he insisted on showing me all the features of his car, which included a hot stone shiatsu massage function. I would have probably really enjoyed this if it weren't for my motion sickness that kicked in shortly after hitting the highway. It was only a dull inkling of nausea at this point so I simply opted to not have the massage and thought it would have corrective actions. But then we went for a leisurely drive around his neighborhood which was all rolling hills and country lanes, dotted with fence posts and horses. It was gorgeous romantic scenery but now my mild nausea was turning into an excessive desire to projectile vomit.

I suggested it was best to do something that required my feet to be firmly on the ground and not cruising in a very smooth German automobile. He pulled in to his new home, a small estate he had recently purchased due to the collapse of his marriage and eventual sale of his previous luxury mansion. It was an absolutely gorgeous home surrounded by acres of open land. He was in the midst of renovations and had made subtle comments of the vast downsizing he had to do. I thought his current new home was absolutely spectacular all the same.

After about 20 minutes he pulled his helicopter out from under covered storage and began to prep it for flight. He quickly ran through the obligatory pre-flight inspection of the propellers and phalanges and whatever other things make up a helicopter. He suited me up with headgear and a microphone and explained how to use it. I jumped in the passenger side and he fired up the blades, and slowly but surely we began to rise up in the air. Legitimately this was the first time I had even been in a helicopter and it was really an amazing experience. It didn't even feel like we were off the ground, but looking out the windows reminded me that we were. He put on some satellite radio station that was all romantic music that streamed into my headset and I tried hard not to giggle at hearing Kenny G, Whitney Houston and Boys II Men wailing in my ears.

He was very much a gentleman and absolutely treated me as such. He took us overhead through the sky above his previous estate and then I understood why the casual remarks for downsizing. Even from a few thousand feet in the air the complex was massive, complete with riding arena for horses. We had made it about 10 minutes further in when the nausea from the car ride had settled back and I was beginning to enjoy the ride a lot less. We were above the valley when I embarrassingly said I did not want to vomit in his pretty helicopter so we had better go home. The disappointment could have moved the earth but he obliged regardless.

He told me once we had landed that he had actually planned on whisking us off to a rather famous winery and spa that he had a hand in creating for some lavish wining and dining. We spent the rest of the day simply talking on a couch in his luxury rental (as his new home was being renovated and clearly unlivable) which turned out to be a very relaxing and lovely occasion. I am so glad I didn't end up throwing up in his helicopter, but I had really wanted the opportunity to learn more about how one flies one of those winged contraptions. It is on my to-do list to learn how to fly a plane one day in the future when I have a little more time, so perhaps that will parlay in to more time in a chopper.

We did not end up pursing any kind of romance, and whether or not it had to do with the vomit I will never be sure, but we do remain friends. I have dated a few people since "Helicopter guy". Most have been very short-lived due to completely opposing world and self-views and would be enough comical fodder for its' own book. The majority of people looking for love didn't seem to be really ready to be honest, but I understand more than many the desire to be coupled with someone for the affirmation and acceptance when newly alone. After you feel like you have lost an appendage when you have lost your partner, it is only natural to want to replace that empty space with someone or something.

Some of the people I have met through awkward attempts at dating have gone on to be good friends. One even stepped in to take my dog's lifeless body to an all-hours animal hospital for cremation after she died, and for that I will be eternally grateful even if we do have very different political views. I am getting better at identifying those people I definitely do not want to spend time with and I see that as a very healthy thing to do considering my tendency to want to fix any bird with a broken wing. I also realize that I deserve to give myself time and appreciate who I was as a person, in all of my splendour, before seeking out someone else. I no longer want someone to complete me, because I have chosen to do that for myself.

CHAPTER FOURTEEN
DESTINATION: CHANGE

My life has taken unexpected turns and twists, far more than I ever thought bearable. It has broken me down to basic parts and caused me to re-evaluate what my values, wishes and dreams truly are. I have suffered loss upon loss and found it extraordinarily difficult to find the strength to pick up the pieces, but I have managed somehow to continue moving forward. It was only when I became frustrated with the way my life was going that I began to seek out change.

I had two main sources of income at the time, one in adult education and another was a small home business I had been working on in the community. They were decent enough ways to spend my time and I tried very hard to do the best I could with them. I wanted more from them though because I saw the potential they could have been, but due to reasons beyond my control I outgrew the space they held for me and it wasn't until I started feeling this vague sense of wanting "more", that I started wondering what that would look like. I started by listing all of the things that I wanted to change and what new achievements would look like.

I wanted a better job first and foremost, one that was more satisfying and used more than just one of my skills because I didn't want to feel that I was stuck spinning my wheels, not moving in any direction personally or professionally. I was tired of the frustration I felt living a life that I felt was just mediocre, with no new ideas or challenges or advancement. The hours were not consistent and I had reached the limits of my pay grade and position responsibilities. I didn't feel fulfilled in any aspect, and I clearly identified that I needed to start seeking out other opportunities.

The first thing I decided I wanted was stability, both financially and logistically, knowing where I would be and what I would be doing. Then I started making notes about what stability meant to me. I started to act and began creating lists and vision boards of all the things I wanted to achieve in plain and certain terms. I wanted to do things that had honest purpose and I wanted more financial security. I wanted to be able to afford to live   in a newer home that wasn't lopsided and full of mold. I wanted to feel like I was doing things that got me closer to who I really was, instead of simply treading water or having four spinning wheels just to pay the bills.

I also started reading books and taking courses in Psychology and Medical Anthropology in an attempt to understand myself and others. I started writing and painting acrylics again and attending ballet lessons. I threw myself into community events that had special meaning to me, and I started sending resumes out to everyone and anyone who may have been looking for an enthusiastic employee within the fields I wanted to pursue. I set up online accounts with every major recruitment firm and employment site. I started networking both in person and in the digital realm. I put myself out in to my local community and abroad as much as I could and I began to compile what my personal vision was for my professional goals. I started focusing on the life I wanted instead of the life I found myself in presently. I forgave myself for every wrong thing I had ever done, and for every situation I found myself to be a part of up until then. I let it all go so I could focus on where I wanted to be.

I spent months doing everything I possibly could to open more doors for myself, and to start giving back to others and in turn myself. The connections I was making locally as well as digitally started to lead to other connections. I applied for more casual work with an organization I felt had values that were aligned with my own, and I did everything I could to bring my enthusiasm and work ethic to every shift I took there. It took months, but eventually I did land a new permanent position that happened to be exactly what I set out to achieve for the company whose values aligned with mine that also gave me the opportunity to work with truly marvellous co-workers and managers. I moved to a new city and found a gorgeous rental condominium in an upscale neighborhood.

The physical move itself went incredibly smoothly. My family and friends completely supported the new opportunity in a way that I never had experienced previously with any of my other past decisions or transitions. Friends offered to help move when I mentioned I was having trouble arranging transportation and other friends I hadn't heard from in over 20 years agreed to help unpack just because I asked. I was able to line up a family doctor who was accepting patients and who happened to be in the same neighborhood, and a dentist who was wise and fair. I was fortunate to get my son into one of the best schools in town even though it was through a lottery system and I have found the opportunity to be a soccer and tennis mom.

I have had the support of my friends who I am sure are quite bored with hearing me go through months of crying and begging for things to get better. Luckily for them I've finished processing the pain of my past, even though it needed to be done. I had to learn to grow as a person and to strip myself down in order to see what I was really made of. I have conquered crushing anxiety and feelings of hopelessness. I have learned how to let go of that which I cannot control and I have learned I am worthy of love and respect and that I am capable of a great many things. I have learned that I actually have the ability to make my life as great as I would like it to be and I had to stop pressing limitations on myself that others imposed on me.

I have watched my dog die under a car, decidedly the most horrific thing to have to witness, and had my son be ill in hospital days after the dog accident--surprising how alike appendicitis and atypical pneumonia present-- all while holding down a full-time job and moving along in daily life. I have learned how to deal with loss upon loss and I have learned how to fail at my current job in a way that allows for my own growth and that of those around me. I have learned the importance of allowing myself to truly grieve the sad and mournful experiences in my life, and to give myself time and grace to feel the pain until it becomes less hard to bear. I have learned the importance of giving back to my friends, family and community. I have learned I actually have quite a bit of experience with life and am finding it easy and exhilarating to have conversations with people who are going through crises of many kinds. By giving back to people who now suffer, it strengthens who I am. When I hold someone else up, it helps me to hold myself up.

It makes sense to me to actually have conversations with people who are moving through experiences that we all would find difficult. I don't have all the experience and I sure don't have all the answers, but I have found that most people are not looking for answers from other people; they just need an understanding ear and eventually a hand to be shown how to find the answers within themselves. The first time someone asked me what I wanted out of life I had no idea how to answer that. I did not know how I wanted to live, I did not know what I wanted to do, or even where I wanted to go or what I wanted to achieve. So of course it was nearly impossible for me to attain any goals if I hadn't actually outlined what I wanted to reach. So often we get overwhelmed with the thought that this means we need to "figure it all out" but we don't, all we need is to ask ourselves "what does my ultimate life look like?" And go one step at a time in that direction.

I do not want to continually low-ball myself out of places to live or employment opportunities or relationship matches. I do not want to always dumb myself down for those I had connections with just as much as I do not want to find myself in relationships that were not serving me the way I deserved to be served. And I want to be able to give back service to all of the people that have helped me in some way, shape or form over the years. I am not a woman of means by any measure but I give what I can to who I can and when I can. I have stopped settling for the bare minimum because I truly believe that my worth far exceeds mediocrity. I want to get comfortable with pure gratitude to those who mean the most to me. I openly tell my friends how much they mean to me and I create gatherings in spaces that are safe in order to share my love for them. I also recognize when I have achieved something I set out to do, even if it is a little achievement like surviving a work week or completing a mundane task. I am doing it. I am completing it, and I am doing it as best I can even if it is not exactly what I want to be doing.

I give back to my community by volunteering for events and neighborhood working groups. I have started carrying around preloaded gift cards in my purse so I can freely give to those who look as though they find themselves in a tough spot in their life. Every life matters, and every person matters. I almost burst into tears every time I provide someone impoverished with a gift card because it is something so simple yet so impactful. I am not handing out hundred or thousand dollar bills but a hot coffee on a cold night to one disadvantaged person can mean the difference between holding on (or not) for another day because someone showed them there was still good in the world. And I thank them for being there and contributing to my day by allowing me to give them back a small gift of gratitude that I have received from someone else in some other form.

That we think of others who are not doing as well financially as ourselves as being of lesser value is an incredibly incomplete world view. For instance, I have found myself with little money and little direction at many times in my life and I have been fortunate enough to have all kinds of people I never would have expected give to me exactly what I needed, when I needed it most. It is my duty to serve and to give back to the world that has picked me up every time I have fallen. I do not do any of these things for any other reason but because I can. I have the ability to do small things for other people and I am going to change the world one life at a time, and the first step has been with changing my own life. I have unlimited potential to be the person I have always been destined to be but never had the courage to embrace.

I have always known I would one day help make this world to be a better place but I had no idea what that would look like or that it all had to begin with me. I thought until recently that if I just gave all of myself to others, if I helped everyone but myself, then that would be the change the world needed. As I have discovered, it starts with me having to make myself better in order to make the world better. I had to first validate my worth in order to really make an impact on others. By acknowledging the value in myself, I am also acknowledging the value in every other person I come in to contact with and everyone they know until infinity. Because I am you, in a sense.

We are all human beings, we are all humans doing to ourselves and each other, and we are all humans learning. My sincere hope is that by saving one life, even if at first it is only my own, is that it will encourage other people to save one life as well. If we all just save one life by focusing on how we can make that life better by being honest with ourselves and with others then we will naturally affect everyone around us in a positive, respectful manner. If we focus on understanding ourselves we may find it easier to understand others and we may find it easier to be gentle with ourselves as well as with others. We are hard on others when we are hard on ourselves, and we cause hurt in others when we hurt in our own self.

I am now a whole being because I have wanted that to be my ultimate goal and I am comfortable in solitude as well as in a crowded room. I am still judged, and I still make mistakes, and I am getting stronger in understanding my own strength to know that I honor myself and my own worth. I now have a stable job that has thankfully offered me more support that I ever could have imagined in my many times of need. Single parenting is hard but I am doing it, and I am actually doing pretty well at it by my own standards. My son has adapted wonderfully to life in a new city and its challenges thus far, and is settling in at school. I am happily rocking the PTA. I have volunteered to DJ at school dances, coach sports for my son's teams, and offer to do mundane tasks like check for lice or hand out lunches.

Sometimes we can get overwhelmed by the thought that we have to save everything, change the whole world or figure all of the worlds issues out all at once, but we don't. It starts with one simple step of recognizing our own worth for to start changing the world one self at a time. We must stop thinking one person can't make a difference, or that the world cannot be made a better place just because a lot needs to improve. Start with the simple step of improving yourself. Just think, if we all focused inside even for five minutes a day it would have an immeasurably positive impact on those around us because we wouldn't be seeking to cause chaos to distract ourselves from our own minds. Despite having lived one very full and challenging life, I am still vivacious and bursting with enthusiasm for new challenges and new experiences. I have, however, changed my focus from the broader sense of having challenging life experiences to a more concentrated thinking that I would prefer my future challenges to not be challenges at all and to definitely not be as soul-destroying as in the past.

It almost feels as though my vivacity and voracity for joy are more honest now because I know I have lived the life I have and I have survived it. I do not live with a chip on my shoulder or lead in my shoes. I do, however, take my sweet time in getting to know anyone I bring near to my life. I look forward to opportunities to grow and learn and become the newer, more enlightened me. I look for the path of least resistance and I ask for help from others in order to get there. I will still most likely trip on my own feet, or laugh at inappropriate times and say the wrong thing occasionally, but at least I am more comfortable in my own skin while doing these things.

I do not want to struggle to make something work that simply does not work, whether it be personally or professionally. We are all taught to fight and struggle and endure but in the wrong sense. We do need to fight, struggle and endure to clarify what our goals are, we need to be able to clearly identify what it is we seek but we do not need to do it by repeatedly banging our head against a wall to get there. I am a firm believer that when things are meant to happen or when the path is the true way to go, everything along that path seems to glide along effortlessly. I don't focus on how I am going to accomplish every little step towards my goal, I only maintain what the goal looks, feels, sounds and tastes like. I move one step at a time, keeping the vision in my mind all the time of where it is I want to reach and I do this while being the best self I can be. I am human and I no longer need to be anybody else's image of perfection, I only need to measure up to my own imperfect self. And I only need to know what my own vision and destination is.

I encourage others to give to themselves as much as they give back to their own communities. One way to give to yourself is by writing a list of all the ways that you are an amazing person. Identify specifically what you have to give to yourself before you begin giving all of that to someone else. Once you identify all the gifts you have and all of the resources you can provide, it will make it easier to start giving outwardly. If that means donating a few cans to the food bank, handing out granola bars to those who hold cardboard signs at intersections, or spending leisure time sitting with a lonely person of any age, then that is enough. Even now when I have a moment of doubt I write down all the reasons why I have no choice but to be successful, and all the skills and attributes I have that serve a purpose of good to myself and others. As long as my intentions are good and giving to myself and others, they will be honored. If you have been blessed with the ability and the means to contribute to someone else's wellbeing in any way, find a way to give that is meaningful to you and you will be exponentially rewarded. There is not one person on the planet that does not require some kind of human interaction, some kind of acceptance or some kind of forgiveness and we all have these things in us to give--especially when it is with ourselves that we need to practice these things. If you find that you are in need of human interaction, acceptance or forgiveness then ask yourself for it and ask others to participate in it. People are listening and will respond.

I am not asking anyone to singlehandedly reforest the planet, stop animal cruelty, save the ice caps and end all wars simultaneously. I just want everyone to ask themselves "What do I want out of life today, in this moment?" Only when we envision our goal can we start moving, one foot in front of the other to get there. It may take you down some unexpected paths and it may take you through the bog of stench and pain and sorrow before you get there, but I promise if you continue to see in your brain what your destination looks like, you will get there, and it will be more amazing than you could have ever imagined because you got there.

I knew from a young age that I wanted to travel the world, but I had no idea how that would happen or how to start that process. But I continued to dream and even thought it came from a place of pain and discomfort it did eventually happen, and will happen more and more in my future because I know it is what brings me joy. We are all global citizens, we are all on this earth to live, grow connect and give. For anyone struggling to make sense of where they find themselves at the current moment, there is always a way to find meaning and it all starts with the desire to start the search. There is always someone willing to help and there is always a tomorrow to continue to put one foot in front of the other. Everyone needs to hear that someone does not want them to give up. And I am willing to be that person for anyone. From the bottom of my heart, no matter how hard life tries to throw any of us a curve ball in order to teach us one of its fabulous lessons, do not give up. Keep going and ask for help. There is more than enough of exactly what you seek as long as the intention is pure, honest and good. You just have to start looking for it. Change happens regardless of whether we actively seek it out, and when we are not actively seeking it out we are forced to change regardless of whether we are ready for it or not. Change is hard and often times painful but it is always for the better even when it feels like it is destroying us.

When I was younger, before any of this life happened, I thought I would be blissfully married and have three kids before what I thought would be the unbelievably old age of 25. I saw this as my goal because I had no idea what else to look for. Society and my surroundings had shown me that I was a woman and should therefore be happy with being a mother, period, as though that was all I could be. I had no idea what I wanted to do, who I wanted to be, or even what I wanted to be good at. I was so lost for so long and I never in a million years would have expected I would have actually travelled the world more than once, met amazing people, had crazy adventures, been married and divorced or have a child out of wedlock. I would never have pictured myself on a hillside in Morocco or wandering a night market eating food from carts in Taiwan.

I would not have had the opportunity to feel the deep sorrow or the elated joy of the path my life has taken me on. I would not have been given the opportunity to find the strength within myself to figure out how to land on my own two feet and accomplish exactly what I desired to do, once I figured out where my desire and intent lay. I am so grateful for all the time spent in far-off countries trying to speak foreign languages and the personal struggles that followed. For all the opportunity I have had and all that lay in front of me, I am truly and humbly grateful. I have struggled, most times because I didn't realize I didn't have to struggle to get to where I am today. I have struggled because deep down I wanted to understand who I was, what my life would be about and how I could parlay that to improving the lives of others. Life can be really good, if one chooses to make it really good. Conversely if we think our lives are terrible, we can't be all that surprised when we only manage to see the ways in which it is not going well.

We are all dealt extremely terrible hands from time to time and there is no getting around it, whether it is a failed marriage, a lost job, an incomplete degree, physical or emotional illness, death, a drug addiction, infidelity, or feeling like a failure as a parent. None of this makes anyone a person of less worth. Maybe you've had the unfortunate experience to have lost everything all at the same time. I am genuinely sorry for all you have been through, all of it, every single moment that has sucked so incredibly badly. But here I am, and here you are. You are still standing, or sitting, or laying in the fetal position reading my words. Once the dust settles and the shock and sorrow of whatever the situation may be has passed, there is always the option to pick yourself up when you are ready, and keep going and ask someone for help. Maybe you also need medical or spiritual intervention, so go get it. There is no shame in asking or needing help, and there are people that are more than happy to be there for everyone. There is enough of whatever we seek, so we just have to start looking. Take the time to grieve the loss, feel all the feelings and make all the poor choices we make following a loss. Don't even feel bad about all the ways you have failed. Embrace that your life has completely exploded and then outline where your next destination will be, even if it is somewhere intermediary like making it through today, and put one foot in front of the other. In fact when we meet, let's high five at how epically our lives have shattered in to a million pieces on the floor, like iridescent glitter exploding from a golden unicorn horn and then keep moving forward. Let us triumph in our epic losses and acknowledge that they have served the purpose to propel us forward to somewhere better. We survived a battle, we are warriors together and we will keep going. The time will come when you will be ready to push forward even if you feel like that time is an impossible place in the future. It will come, so be patient and start looking forward.

Looking back it is as though my life has been SCREAMING at me to try and get me to wake up from a life of mediocrity that was never destined for me. I had resigned myself to thinking that I should just be happy with whatever I could get even when it was neither fulfilling nor full of the splendor that I knew deep down I was capable of. Finally I have woken up with the sun almost blinding me with its beautiful and abundant light and I am living the life that I want, regardless of the life others may want me to live, and regardless of how others would like me to suffer because they chose to. I have made amazingly poor mistakes, I have lowballed myself repeatedly and I have unsurprisingly been left in shambles over and over again. I was choosing that life, I chose those outcomes on a deeper level. I had to go and touch the bottom of my own egocentric abyss before I could feel the thirst for greatness I now have.

I am no longer afraid to fail by someone else's measure of me, because I have already succeeded by my own standards. I am doing better than I ever have before now because I am doing what I can to live the life of purpose that I want for myself and my child. I didn't speak much about my life with many people before this. A few select close friends may know many time periods within this book but I'm not terribly certain I have ever told the whole story in as much detail as this before now. I was honestly too full of shame and guilt for every step and misstep of my journey. I didn't think any portion of my adventure was of any value, that no one would want to hear about it. I honestly believed I had nothing to give that was of any worth. And as terribly sad as that may sound it at least made me break out of my own thinking in order to come out of my cocoon to become a terribly cliché Monarch butterfly covered in fairy dust, brimming with wonderment. I am not mediocre, and I am not worthless. I have flaws but I am not garbage. I am not a bazillionaire but I don't have to be in order to feel like I have succeeded.

I like nice things, but I don't need them in order to feel that I am admired by my peers. I feel there is much more value in being respected because I conduct myself with honesty and compassion. I value myself for my efforts in wanting to improve the lives of those around me, whether they are co-workers, colleagues, friends or family. I may have failed in the past, and may continue to fail at doing things in the future, but I am at least trying to learn and grow and be a better person for myself. I have shifted my thinking to be focused on the power of the possible, because I do believe we are able to do much more good than we give ourselves credit for. I have learned to stop focusing on all the reasons why something might not work out because equally as much as something might possibly not work out, it could also work in my favor. There is every reason in the world why I will attain the goals I set in front of myself. I have been able to survive poor relationships, unstable jobs, volatile travelling situations, poverty, loss and personal strife. I have been able to find the courage, strength and skill to forward myself when there was every reason for me to fail. I have carved out a life for myself and my son because I wanted to and because I had to. By my own merits I am succeeding.

Nothing needs to seem impossible, but sometimes we need to re-evaluate the path we are on to see whether it is serving us at our core or not. Things have seemed most difficult for me when I have felt I am wearing someone else's hopes, dreams and desires. Situations have seemed nearly impossible when I am dead set in trying to convince myself the life that I have created for myself is what I want and need when it really doesn't serve my heart or soul. I have had crushing anxiety and have had to work hard at rewiring my brain to not dive headlong into dark disquiet. I have lived in lies and half-truths others have fed me, and I have believed my own. But given enough time, the truth always comes to light.

The truth of who I am, in my brilliance and my liabilities is beginning to come through because I have finally made the decision to honor everything I am, and everything I am not. I still feel all the feelings, but I am learning how to better cope with them. I still am presented with challenges that leave me tense and questioning my abilities, but I am still managing to stumble through them and figure out the lessons within them. I have humbled myself enough to accept that I am worthy of all the good coming my way and not just the bad.

I no longer hold myself to a higher regard than I can live up to, and I try to make every day an improvement on the last. I am finding the strength to allow others to accept responsibility for their own actions regardless of how badly I want them to succeed. If my co-workers, friends or lovers fail I am trying to emphasize that it is not because they are not the person I wanted them to be, but because they too are human, and they too are on their own quest to figure out where to build their nest in the universe. I have had to plan a redesign for my life on more than one occasion, and it has not been done easily. I have somehow managed to continue living, growing and learning despite all of it. The way I view myself and my thoughts has changed dramatically recently. I have a better understanding of how to start planning the life I want, need and desire and am slowly taking steps in the right direction.

It is our thoughts that preconceive the notions of our future selves and I have learned the hard way how painfully necessary it is to completely break away from negative thinking to pursue my own purpose, which is to share my life and my joy. I have always wanted to share my honest life with someone else in more depth than passing conversation. So if only one person manages to read through this entire thing and gains something positive from it, then it has truly lived its purpose. My path has twisted and turned and been full of spontaneous joy and adventure as well as drawn out sorrow and self-discovery. I am doing many things that just a few short years or even months ago I never would have imagined to be possibilities for me. I have done it largely without financial help or the safety net of a secure partner. I have done it while simultaneously continuing my own fight for redemption and have shown myself I am capable even when others would try their best to convince me otherwise.

There will always be those who want to impede the success of others, but I continue to plan and dream and plot my course amongst the stars because I can see for the first time the unlimited potential you and I all possess. It will not be straightforward, and at times everything will seem less certain and less hopeful, but everything I am doing is advancing me forward and not back. I believe we are capable of amazing things, whether it be tackling homelessness or ending wars or simply standing up to embrace our own greatness. It all begins with our selves first. Making the conscious decision to improve our own lives in the little ways which we can do on a daily basis, making the conscious decision to figure out how we can find satiety internally for ourselves instead of in external sources. Because focusing our attention on what we can do to make ourselves less angry, more productive, healthier, more contented with ourselves will benefit every other person we are connected with in the long run. And I am in it for the long run. I cannot and will not run away from the things I fear any longer--except maybe guns and bombs, and I'd probably still run away from those.

It all began because I started to move towards healing myself, finding myself, searching for my present and my future self and letting go of my past. It started when I began to recognize my own value, my own worth, my own talents and my own desires. I have begun a movement all for myself that celebrates failures and losses and adapts to embrace the strengths and opportunities that they present. I try to steer my thoughts always to finding the truth and the opportunity even when faced with seemingly insurmountable challenges and I embrace those who have sought out these opportunities in their struggles as well. I have begun to believe in the infinite power of possibility that reaches in every direction. I am beginning a movement for my one life, because I am worth saving. It has not been easy to save my own life, but I have done it--one step at a time. And if I, with my awkward clumsiness and ability to repeatedly make all the wrong decisions can do it, anyone can.

Every time I have made a goal for myself, however haphazardly, it has always been achieved. Now I focus on clearly thinking out those goals and the motivation behind them so that when I do achieve them they are all propelling me forward instead of in reverse. Whether it is to travel, for family, for success, for financial reasons, in search of peace and enlightenment, anything is attainable. I train myself to not worry about every minute of the journey to the destination, I figure out what that destination looks like, smells like and tastes like. How do I feel once at this destination, what kind of people are with me and what kinds of things do I do when I get there? What kind of car am I driving, or maybe I am on a boat or a plane or even a kayak. What kind of food am I eating and what kind of friends am I making while there? Slowly but surely things will start rearranging and moving towards these goals if they are clearly laid out. It will happen, of that I am sure--I remind myself to be patient and keep searching. I try to keep digging and keep whittling down my vision of where the next destination will be, as the clearer I become, the easier it will be to attain. The further along I get on this new, clearer path I have begun to carve out for myself the more I am beginning to notice all these amazing things that are starting to happen. I have had the most random conversations that happened to be exactly the right words at exactly the right times. I actually know, beyond the shadow of a doubt that I am at exactly the right spot along my one life movement and that feels very good. Finally.

When you decide you are ready and willing to begin your journey, your life, your quest, let me know. And I will help however I can with whomever I can simply because I am able to. Your life is worth it.

All of this began with one life, my own--because it all begins with one life and I am living The One Life Movement, one step at a time.

CONNECT

One of the most important underlying themes of this book is the importance of getting connected. Not for any other sake than to come to the realization that our personal health is so dependent on our connectivity with those around us. The more we come to understand how we all impact each other, the greater and healthier we will become. The healthier we become ourselves, the healthier we will become as a society, and as a world in general.

When we start respecting ourselves again, we will respect each other, and the world around us. We will stop wasting, we will stop seeking to destroy ourselves and each other and we will begin truly living. I am not a perfect human being, but I am trying to do my best to demonstrate and honor the values that I hold dear so that all of the people around me can profit intrinsically from that. I encourage everyone to join in community groups, volunteer activities, sports programs or anything that helps you to feel useful and connected to humans, animals or the outdoors.

Become an active participant of The One Life Movement online and offline, because one person really can make all the difference in the world.

Visit www.theonelifemovement.com for more information, photos, experiences and ideas.

Twitter @TheOneLifeMVMNT

Facebook page: www.facebook.com/theonelifemovement

I am ever so grateful that you took the time to read this book, if you enjoyed it, please take a moment to leave a review at your favorite retailer.

ABOUT THE AUTHOR

Kimberley Dickinson is a mother, a nurse, a leader, a manager, an educator, a cook, an entrepreneur, a writer, a sports enthusiast and not too bad at creativity. It has taken years for her to begin her own movement towards wholeness and understanding, and she is completely delighted that she has survived all of the lessons that have thrown themselves along her path to purpose.

She lives on a beautiful island in Western Canada where she spends her time attempting to inspire her son, those around her and anyone willing to connect. She believes in promoting anything positive, including the accolades of those around her, for she understands by lifting up others to their goodness she is also lifting herself up to the light, to the strength and to ultimate truth. She still makes mistakes, but she is still trying. To continue to search for new ways of finding the opportunity in anything, to continue to break through fear and misunderstanding are to live in this new era.

She is currently working on the next book in The One Life Movement series and regularly maintains the website that ties it all together. www.theonelifemovement.com Join her on various social media sites, send an email or message and get in touch!

"We create what we are, and we are what we create. We have the power to make great change, and I plan on doing everything I can to contribute to a healthy, sustainable, kind future for everyone I have ever met, especially those that never thought in a million years they would have the courage to find their own voice. Any time someone tells you anything can't be done, do not listen. Rephrase it as everything is possible, I just have to find a way, and there is always a way."

–Kimberley Dickinson